IEE

Guidance Note 1
Selection & Erection
(Including Amd No 2 : 1997)

BS 7671 : 1992 Requirements for Electrical Installations

Published by: The Institution of Electrical Engineers, Savoy Place, LONDON, United Kingdom WC2R 0BL.

Issued August 1992
Reprinted April 1993, with amendments
2nd edition incorporating Amendment No 1 to BS 7671 — May 1996
3rd edition incorporating Amendment No 2 to BS 7671 — Dec 1999

Copies may be obtained from:

The Institution of Electrical Engineers
PO Box 96, STEVENAGE,
United Kingdom. SG1 2SD

Tel: +44 (0)1438 767 328
Fax: +44 (0)1438 742 792
Email: sales@iee.org.uk
http://www.iee.org.uk/publish/ books/wireregs.html

ISBN 0 85296 954 6,1998

Contents

Acknowledgements

References to British Standards, CENELEC Harmonisation Documents and International Electrotechnical Committee standards are made with the kind permission of BSI. Complete copies can be obtained by post from:

BSI Customer Services
389 Chiswick High Road
London
W4 4AL

Tel:	General Switchboard:	0208 996 9000
	For ordering:	0208 996 7000
	For information or advice:	0208 996 7111
	For membership:	0208 996 7002
Fax:	For orders:	0208 996 7001
	For information or advice:	0208 996 7048

BSI operates an export advisory service — Technical Help to Exporters — which can advise on the requirements of foreign laws and standards. The BSI also maintains stocks of international and foreign standards, with many English translations.

Up-to-date information on BSI standards can be obtained from the BSI website http://www.bsi.org.uk .

Illustrations and information of the forms of separation of LV switchgear were provided by The Low Voltage Distribution Switchboard Division of EIEMA. The EIEMA booklet 'Guide to forms of separation' can be obtained from:

EIEMA
Westminster Tower
3 Albert Embankment
London
SE1 7SL.

Extracts from NHBC Standards, Chapter 8.1, for internal services, on the requirement for notches and holes in joists are reproduced with permission from The National House Building Council (NHBC), Buildmark House, Chiltern Avenue, Amersham, Bucks, HP6 5AP.

Extracts from the LDSA document 'Fire safety guide No 1' on Section 20 buildings are reproduced with permission from The London District Surveyors Association, PO Box 15, London, SW6 3TU.

Guidance on cable separation in Appendix J was reproduced from The Electrical Contractors Association document 'Recommended cable separations to achieve electromagnetic compatibility (EMC) in buildings' with the permission of the ECA, ESCA House, 34 Palace Court, London, W2 4HY.

The table on colour identification of buried services is reproduced with the permission of The National Joint Utilities Group, 30 Millbank, London, SW1P 4RD.

The tables on cable references in Appendix F were reproduced with permission of Anixter (UK) Ltd, 15 Chesford Grange, Woolston, Warrington, Cheshire, WA1 4RQ.

Copies of Health and Safety Executive documents and approved codes of practice (ACOP) can be obtained from:

HSE Books
P O Box 1999
Sudbury
Suffolk CO10 6FS

Tel: 01787 881165
Fax: 01787 313995

Health and Safety Executive telephone enquiries can be made to:

HSE Info Line on : 0541 545500

Fax and postal enquiries can be made to:

HSE Information Centre
Broad Lane
Sheffield S3 7HQ

Fax: 0114 289 2333.

Preface

This Guidance Note is part of a series issued by the Wiring Regulations Policy Committee of the Institution of Electrical Engineers to enlarge upon and amplify some of the requirements in BS 7671 : 1992, (formerly the Sixteenth Edition of the IEE Wiring Regulations).

The scope generally follows that of the Regulations and the principal Section numbers are shown on the left. The relevant Regulations and Appendices are noted in the right hand margin. Some Guidance Notes also contain material not included in the BS 7671 but which was included in earlier editions. All of the Guidance Notes contain reference to other relevant sources of information

Electrical installations in the United Kingdom which comply with BS 7671 are likely to satisfy the relevant aspects of Statutory Regulations such as the Electricity at Work Regulations 1989, but this cannot be guaranteed. It is stressed that it is essential to establish which Statutory and other Regulations apply and to install accordingly. For example an installation in premises subject to licensing may have requirements different from, or additional to, BS 7671 which will take precedence.

The 16th Edition of the IEE Regulations was adopted in 1992 as British Standard BS 7671 : 1992. Throughout the text of this Guidance Note all references to the IEE Regulations have been replaced with BS 7671. This reference must be taken to incorporate Amendment No 1, 1994 (AMD 8536) and the subsequently published Corrigenda (AMD 8754) and Amendment No. 2, 1997 (AMD 9781).

Introduction

This Guidance Note is concerned with Part 5 — Selection and Erection of Equipment.

Neither BS 7671 nor the Guidance Notes are design guides. It is essential to prepare a full design and specification prior to commencement or alteration of an electrical installation. Compliance with the relevant standards should be required. 120-01-02

The design and specification should set out the requirements and provide sufficient information to enable competent persons to carry out the installation and to commission it. A good specification must include a description of how the system is to operate and all the design and operational parameters. It must provide for all the commissioning procedures that will be required and for the provision of adequate information to the user. This will be by means of an operational manual or schedule, and 'as fitted' drawings if necessary. 514-01 514-09

It must be noted that it is a matter of contract as to which person or organisation is responsible for the production of the parts of the design, specification installation and any operational information. The persons or organisations who may be concerned in the preparation of the works include:

The Designer
The Planning Supervisor
The Installer (or Contractor)
The Supplier of Electricity
The installation Owner (Client) and/or User
The Architect
The Fire Prevention Officer
Any Regulatory Authority
Any Licensing Authority
The Health and Safety Executive

In producing the design and specification advice should be sought from the installation owner and/or user as to the intended use. Often, as in a speculative building, the intended use is unknown. In such cases the specification and/or the operational manual must set out the basis of use for which the installation is suitable.

Precise details of each item of equipment specified 120-04
should be obtained from the manufacturer and/or 120-05
supplier and compliance with appropriate standards 511
confirmed.

The operational manual must include a description of how the system as installed is to operate and all commissioning records. The manual should also include manufacturers' technical data for all items of electrical equipment, wiring. switchgear, accessories, etc. and any special instructions that may be needed. The Health and Safety at Work etc. Act 1974 (HSWA) Section 6, and the Construction Design and Management (CDM) Regulations are concerned with the provision of information. Guidance on the preparation of technical manuals is given in BS 4884 (Specification for technical manuals) and BS 4940 (Recommendations for the presentation of technical information about products and services in the construction industry). The size and complexity of the installation will dictate the nature and extent of the manual.

Section 1 — The General Requirements

1.1 General
510

'Equipment' is short for 'electrical equipment' and is defined as: Part 2

Any item for such purposes as generation, conversion, transmission, distribution or utilization of electrical energy, such as machines, transformers, apparatus, measuring instruments, protective devices, wiring systems, accessories, appliances and luminaires.

Installation designers may need advice from the manufacturer as to the suitability of equipment for its intended use. It is not the intention of BS 7671 to stifle innovation or new techniques but the Standard only recognises and considers established materials and techniques. 120-04-01 120-05

1.2 Equipment
120
510

Only installations which are correctly selected and erected comply with BS 7671. It is the responsibility of the designer to ensure that the design may be safe. All departures from BS 7671, although the designer is confident regarding safety, must be recorded in the Completion Certificate.

Chapter 13 of BS 7671 : 'Fundamental requirements for safety' outlines basic requirements. Later Chapters describe in more detail particular means of compliance with Chapter 13. Chapter 13 is normally referred to only where it is intended to adopt a method not recognised in later Chapters. The terms allow for interpretation to suit special cases. Chap 13

The phrase 'so far as is reasonably practicable' is used in several Regulations. It should be borne in mind that methods described in later Chapters are considered reasonably practicable in most circumstances. Unconventional installations may lead to a degree of safety less than that required by Chapter 13. It is unlikely that a designer could then 130-02 130-10 claim compliance with Chapter 13 'so far as is reasonably practicable'.

1.3 Electricity at Work Regulations 1989

The requirements of The Electricity at Work Regulations (1989) are intended to provide for the safety of persons gaining access to or working with electrical equipment.

The Memorandum of Guidance on The Electricity at Work Regulations (1989) (Health and Safety Executive Publication HS(R)25) should be carefully studied and it should be borne in mind that The Electricity at Work Regulations (1989) (The EAW Regulations) apply to designers, installers and users of installations alike.

BS 7671 is intended primarily for designers and installers and thus covers the design of the installation and the selection and erection of electrical equipment. The user has the responsibility for ensuring that equipment is properly operated and maintained when necessary. The installation designer should assess the expected maintenance and the initial design should make provision for maintenance 341-01
to be carried out.

Within The EAW Regulations is the requirement for adequate maintenance to prevent danger, and the memorandum to The EAW Regulations advises that regular inspection of electrical systems, (supplemented by testing as necessary) is an essential part of any preventive maintenance programme. Regular operational functional testing of safety circuits (emergency switching/stopping etc) may be required — because unlike functional circuits they may be infrequently used. Comprehensive records of all inspections and tests should be made, retained and reviewed for any trends that may arise. The IEE Guidance Note No 3 on Inspection and Testing gives guidance on initial and periodic inspection and testing of installations.

1.4 The Construction (Design and Management) Regulations 1994

The Construction (Design and Management) Regulations 1994 (the CDM Regulations) made under the Health and Safety at Work Act 1974, came into force on the 31 March 1995, and were fully implemented on the 1 January 1996. The CDM Regulations place responsibilities on most installation owners and their professional design teams to ensure a continuous consideration of health and safety requirements during the design, construction, and throughout the life of an installation, including maintenance, repair and demolition. This includes design of the electrical installation, and selection and erection of electrical equipment for health and safety in installation and general operation. Design work should take into account the practicalities of installation and adequate installation and operation access and maintenance requirements for all equipment. It is important that all those who can contribute to the health and safety of a construction project, particularly clients (the installation users) and designers, understand what they and others need to do under the CDM Regulations, and discharge their responsibilities accordingly. This will require a radical change in culture for many of the new duty holders.

Designers must consider the need to design in a way which avoids foreseeable risks to health and safety or reduces these risks as far as practicable so that projects they design can be constructed, operated and maintained safely. Where risks cannot altogether be avoided, information on them has to be provided for inclusion in the project health and safety plan. The designer must also co-operate with others to enable all parties to comply with relevant statutory requirements and prohibitions placed on them.

Section 2 — Selection and Erection of Equipment

2.1	**Selection and erection of equipment 510**	Chapter 51 gives the basic common rules to which every installation must comply.	510-01-01
2.2	**Operational conditions and external influences 512**	An assessment by the designer of installation characteristics and conditions will be necessary, including all the requirements of Part 3 of BS 7671. The installation must be designed to be suitable for all the relevant conditions and external influences foreseen, including electricity supply effects or effects on the supply.	Part 3 512 510-01-01 331-01-01
		Appendix 5 of BS 7671 details the system of classification of external influences developed in IEC 60364-3, and the classification is indicated in parts of Chapter 52. This system is not in general use in the UK but certain parts can serve as a reminder of conditions to be considered.	Chap 32 Chap 52 Appx 5
		All equipment must be selected to accommodate the worst foreseeable conditions of service that can be encountered even if such conditions happen rarely.	510-01-01
2.3	**Compliance with standards 511**	BS 7671 recognises equipment which complies with a British Standard or Harmonised Standard appropriate to the intended use of the equipment without further qualification. A Harmonised Standard is defined in Part 2 as 'A standard which has been drawn up by common agreement between national standards bodies notified to the European Commission by all member states and published under national procedures'.	511-01-01 Part 2
		There is a statutory definition of a Harmonised Standard in the Health and Safety Statutory	

Instrument No 3073 (1992) 'The Supply of Machinery (Safety) Regulations 1992'. This states:

'Harmonised standard' means a technical specification adopted by the European Committee for Standardisation (CEN) or the European Committee for Electrotechnical Standardisation(CENELEC) or both, upon a mandate from the Commission in accordance with Council Directive 83/189/EEC of 28 March 1983 laying down a procedure for the provision of information in the field of technical standards and regulations, and of which the reference number is published in the Official Journal of the European Communities.

European Norm EN standards are standards that are required to be adopted by all CENELEC members and to be published with identical text by all members, without any additions, deletions or further technical amendments. Such EN standards then supersede the relevant national standards which are withdrawn to an agreed timescale. They are published as BS EN's in the UK.

Harmonisation Documents (HD's) forming the CENELEC 384 standard are documents based on IEC 60364. Harmonised Standards are standards that have been agreed by all CENELEC members (possibly with national modifications); but a National Standard may have further technical additions (so long as they are not in conflict with the HD — but not deletions — made by the relevant national standards committee. BS 7671 is based on a number of CENELEC HD's within the 384 series (see the Preface of BS 7671 on page viii) with extra specific technical material added.

Equipment is sometimes satisfactory only when used in a particular way or with other matching equipment. Certain equipment complying with a foreign standard may be safe when used, for example, with a foreign wiring system, but may not be safe when used in conjunction with traditional UK 511-01-01
practice. Where equipment complying with a foreign standard based on an IEC standard is specified, the designer or specifier must verify that the equipment is at least as safe as similar equipment complying with the relevant British or Harmonised Standard. It is the designer's responsibility to ensure that equipment the

designer specifies which is not to a British or Harmonised Standard provides the required performance and degree of protection.

Note that BS 7671 does not insist on approval or certification of the equipment to the relevant Standard, but this may be required in some cases by legislation or by the client.

2.4 Operational requirements 512

Equipment of all types must be suitable for its situation and use. 512

Regulation 331-01-01 lists some examples of characteristics which may adversely affect the equipment to be installed or the electricity supply or other services. This list is not exhaustive but shows some details of what the designer should consider. Equipment must operate safely and should be efficient and correctly selected . 331-01-01

The assessment of general characteristics requires a thorough consideration of the purpose of the installation, the building and installation structure and the electricity supply characteristics including voltage, current and frequency, which must be noted during the first part of the design. If, for example, there is doubt that a switch or circuit-breaker can be used with inductive or capacitive circuits (e.g. motors, transformers or fluorescent lighting) advice should be obtained from the manufacturer. 300-01 512-01 512-02 512-03 512-05-01 331-01-01

All equipment must be selected and erected so as to allow safe working, prevent harmful effects to other equipment and not impair the supply arrangements. This includes the consideration of Electromagnetic compatibility (EMC) effects, as well as more straightforward considerations such as loading, current and voltage rating, circuit arrangements, etc. Regulation Group 515-02 specifically requires consideration of EMC when selecting and installing equipment. Both electromagnetic immunity (EMI); the protection of equipment from other electromagnetic influences (Regulation 515-02-01); and electromagnetic emission; the production and emission of electromagnetic interference by the proposed equipment (Regulation 515-02-02); must be considered. This is a specialist application and advice 515-02 from the equipment manufacturer should generally be taken. (See Appendix I for further information on cable separation.)

Harmonics are an aspect of EMC. Harmonic voltages and currents can cause interference with the normal operation of equipment and overload cables in certain cases (see Sections 2.6 and 6.3 also).

Electric motors may have similar power ratings but different applications. Motors for lifts, industrial plant and machinery, propulsion or ventilation will have differing duty cycles, and will usually be the subject of a Standard. The fixed wiring must be able to match the duty cycles of the connected load. Infrequently used motor-driven equipment with brakes will have different demands on the fixed wiring when compared with, say, a hydro extractor which uses 'plugging' as a method of braking the equipment frequently. The demands on the fixed wiring should be established from the manufacturer's installation instructions. These details will form part of the installation manual. (See Guidance Note No 6 for detailed information on cable selection for motor circuits.) 512-04 512 552-01-01

2.5 Identification 514

CABLES

The identification colours for non-flexible cables for fixed wiring are given in Table 51A of BS 7671 — reproduced here as Table 1 for convenience. Table 51A

The Regulations allow the use of flexible cables and flexible cords for fixed wiring, subject to the relevant provisions of the Regulations being met. These provisions are generally considered to relate to physical installation criteria but it should be remembered that the core colours of flexible cables and cords are not the same as those for fixed cables in Table 51A and are laid down in Table 51B. 521-01-04 Table 51A Table 51B

There is no definition of a "fixed" cable and, as the colour coding for flexible cables is by now well understood, there is no need to identify the cores of flexible cables or cords when their colour and function comply with Table 51B - even if they have been fixed. Table 51B

Any other identification of the cores of a flexible cord or cable is a deviation from the Wiring Regulations requirements and should only be done after careful consideration as to whether in the circumstances it is impossible to comply with the requirements, and the level of safety provided is not less than that by compliance.

The notes of Table 51B say that if the blue core is used for another purpose or it is necessary to identify the phases this is to be done by letters i.e. L1, L2 or L3. A blue core may be used as a phase conductor if the circuit does not have a neutral and this should be identified by means of a suitable letter. Table 51A

If after consideration it is decided that a core is to be used for a function for which its colour is not recognised by Table 51B then its function is to be identified by letters. A brown core used as a neutral should be identified with an N and a blue core used as a phase conductor should be identified by L1 or L2 or L3. It is better, however, to select the correct cable and identification colours for an installation. Table 51B

It is a deviation to apply colours to flexible cores to identify them as being used for a purpose other than that which their original colour indicates - if after due consideration colours are applied the deviation must be clearly recorded on the Electrical Installation Certificate and in the construction record documentation.

A green/yellow core is not allowed to be used under any circumstances as anything but a protective conductor.

There is also a current installation trend to supply and install multicore cables complying with HD 324: 1977 that have blue insulation on the neutral conductor. It must be noted that this may comply with Regulation 511-01-01 but not with Table 51A. Such cables should be correctly identified at their terminations. 514-06-01

CIRCUITS

A diagram, chart or table giving details of the circuits is required. These details must be used by the person verifying compliance with BS 7671; for example, details of the selection and characteristics of the protective devices are needed for verifying protection 514-09 413-02-04 against overcurrent and electric shock. The same details will be needed for periodic inspection and testing later on.

A durable copy of the details should be fixed in or near the distribution board which serves the area. 712-01-03(xvii)

For a simple installation the Electrical Installation Certificate, together with the schedule of test results, will meet BS 7671 provided each circuit is identified in the distribution board (see BS 5486 and BS EN 60439). 514-09 741-01-01

SWITCHGEAR

Labelling of switchgear is very important particularly where the route of the final circuit cables is not obvious. If there is a possibility of confusion some reliable means of identification must be clearly visible. It is necessary for the protective devices to be marked so that they can be identified easily by the user.

Fig 1 is a diagram of a typical installation showing the information needed to comply with BS 7671. The method of presentation will depend upon the size and complexity of the installation. 514-09

Complex installations demand greater detail. Details of protective measures and cables should be provided as part of the 'As installed' information. When the occupancy of the premises changes the new occupier should have sufficient information to properly operate the electrical installation. Diagrams, charts, tables and schedules should be kept up to date. Such items are essential aids in the maintenance and periodic inspection and testing of an installation.

PROTECTIVE CONDUCTORS

Protective conductors used for 'earth-free local equipotential bonding' must not be confused with other cables of similar colour coding. Steps, including labelling, should be taken to reduce the risk of a mistake leading to a wrong connection to such bonding. Bonding to earth, which would lead to danger, is not allowed. The same steps apply to 'electrical separation' where again equipotential bonding conductors are not connected to earth. 471-11 471-12 514-13-02

Bonding connections to extraneous-conductive-parts must be clearly labelled. 514-13-01(ii)

NOTICES

The notices called for in Section 514 of BS 7671 are intended to warn skilled and instructed persons against the risk of working on live parts which may have been thought to be isolated, and persons should be made aware of possible dangers. Unskilled persons should not have access to work on electrical equipment. Only three-phase circuits are normally expected to exceed 250 V. BS 7671 does not require external warning notices to indicate equipment voltage levels, but any warning notice provided must be visible before access can be gained to live parts, simultaneously accessible, with more than 250 V between them or to earth. 514

SAFETY SIGNS

Item 9 of Appendix 2 of BS 7671 states that where a pictographic safety sign is used for a caution of risk of electric shock, the Safety Signs Regulations 1996 apply. Risks should be minimised at the design of an installation and warning signs only used where other methods of avoiding danger are not practical to implement. Appx 2

WARNING NOTICES

Table 2 lists the requirements for warning notices and identification labels given in BS 7671. Labels and warning notices should be of a size and type suitable for the location and installed such that they will not be painted over or easily removed or defaced. Labels etc. should be permanently fixed, eg by suitable screws or rivets, or resin glues, taking care not to damage equipment, invalidate IP ratings or block vents. Stick-on labels are not preferred and should only be used where heat or damp is not expected, as they cannot be considered permanent.

PIPES AND DUCTS

BS 1710 : 1984 prescribes a system for the identification of piped services, ducts and electrical conduits by colour. In the system, red is reserved as a safety colour for fire fighting and yellow for warning. Electrical conduits can be identified by completely painting the conduit or painting bands (150 mm in length). The colour for electrical services is orange (06E51 in BS 4800).

Fig 1: **Typical installation**

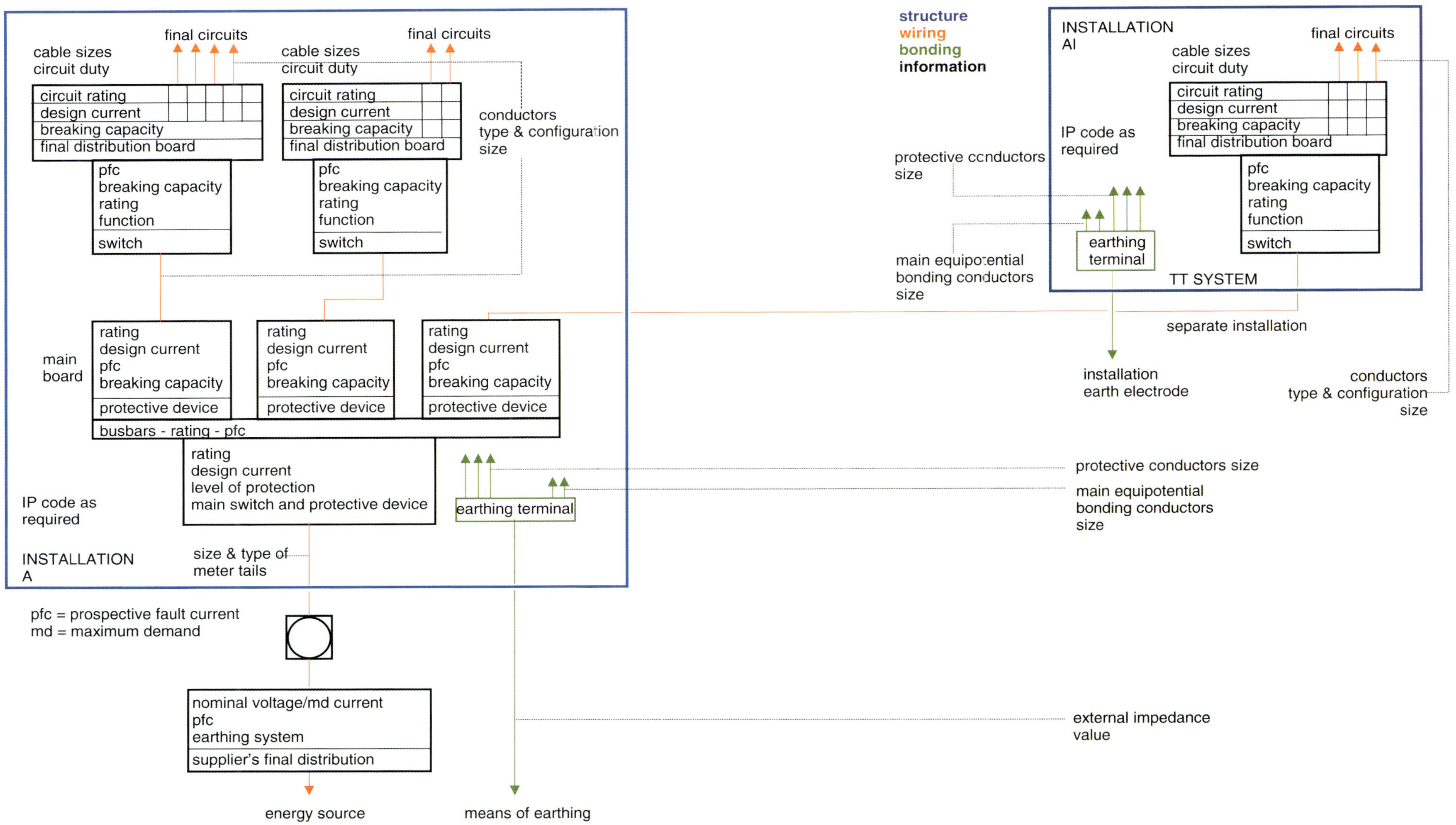

TABLE 1 *(From TABLE 51A of BS 7671)*

Colour identification of cores of non-flexible cables and bare conductors for fixed wiring

Function	Colour identification
Protective (including earthing) conductor and circuits	green-and-yellow
Phase of a.c. single-phase circuit	red (1)
Neutral of a.c. single- or three-phase circuit	black
Phase R of three-phase a.c. circuit	red
Phase Y of three-phase a.c. circuit	yellow
Phase B of three-phase a.c. circuit	blue
Three-wire 460/230 V single-phase a.c. circuits (Centre conductor earthed)	
Outer, phase conductors (either conductor)	red
Centre neutral conductor	black
Two-wire unearthed d.c. circuits	
Positive of d.c. two-wire circuit	red
Negative of d.c. two-wire circuit	black
Three-wire d.c. circuits	
Outer (positive or negative) of d.c. two-wire circuit derived from three-wire system	red
Positive of three-wire d.c. circuit	red
Middle wire of three-wire d.c. circuit (2)	black
Negative of three-wire d.c. circuit	blue
Two-wire earthed d.c. circuits	
positive (of negative earthed) circuit	red
negative (of negative earthed) circuit	black
positive (of positive earthed) circuit	black
negative (of positive earthed) circuit	blue
Functional Earth	cream

Notes:

(1) As alternatives to the use of red, if desired, in large installations, for the supply cables to final distribution boards yellow and blue may also be used. **All single-phase final circuit non-flexible fixed wiring must be red and black.**

(2) Only the middle wire of three-wire circuits may be earthed.

(3) The colours in Table 1 only apply to fixed installations in buildings and not necessarily to installations in factory built equipment which is excluded from BS 7671 (see Regulation 110-04-01). BS EN 60204-1 details colours for the electrical equipment of machines.

TABLE 2

Warning notices and identification labels required by BS 7671

Notice, label or identification	Regulation
Isolating device and its installation or circuit (if not obvious)	461-01-05
Switching device for mechanical maintenance (indication of operation)	462-01-02 537-03-02
Protection by electrical separation	471-12-01
Protection by earth-free local equipotential bonding	471-11-01 514-13-02
Areas reserved for skilled or instructed persons	471-13-03 514-13-02
Fireman's emergency switch	476-03-07 537-04-06
Purpose of switchgear and controlgear (if not obvious)	514-01-01
Conduit, identification, where required	514-02-01
Bare protective conductors	514-03-01
Diagrams, charts etc. for installation information	514-09-01
Marking voltages present on means of access	514-10-01
Voltages exceeding 250 V	514-10-01
Live parts not capable of being isolated by a single device	514-11-01 761-01-03
Periodic inspection and testing (see Regulation 611-04-04)	514-12-01
Quarterly test of a residual current device (see Regulation 611-04-04)	514-12-02
Earthing and equipotential bonding connections	514-13-01 542-03-03
Earth-free equipotential bonding	514-13-02
Cables buried in the ground	522-06-03
Caravan inlet	608-07-03
Caravan instructions for supply and periodic inspection and testing	608-07-05
Identification of highway power supply cables	611-04-03
Highway temporary supply unit (identification by marking etc.)	611-06-02

2.6 Mutual detrimental influence 515

Regulation 515-01 requires that there be no harmful effect between electrical and other installations. The best approach, where practicable, is to arrange that the installations are kept separated. Damage can be caused by such as thermal effects, electrolysis or corrosion. The thermal effects of other installations — such as hot water systems — must be considered and equipment either designed to operate properly at elevated temperatures or reduced temperatures, or be protected from such effects. Electrolysis may result from leakage currents, or from contact between dissimilar metals in damp conditions, while corrosion may result for example in rusting of unprotected steelwork. Action needs to be taken to obviate these risks. 515-01

The introduction of the Electro Magnetic Compatibility (EMC) Regulations places a statutory requirement on designers and constructors to so design/construct electrical equipment and systems so that they do not cause excessive electromagnetic interference (emissions) and are not unduly affected by electromagnetic interference from other electrical equipment or systems. The full implications of such requirements are yet to be understood, and little work has yet been carried out on the implications for electrical installations. Harmonic production and interference, electrostatic discharges, mains-borne signals etc are all types of EM interference to be considered, but the subject is too wide for this Guidance Note and requires specialist advice (see Section 7, and Appendix I for some further information). 515-02

The protection provided should form a prominent feature of the Building Services Manual. Guidance in the preparation of the manual can be found in BS 4884, BS 4940 and Section 6 of the Health and Safety at Work etc. Act 1974 (HSW Act). Manufacturers' leaflets for switchgear, luminaires, electrical equipment, accessories, and any special instructions, should be included.

The Health and Safety at Work etc. Act 1974, the Consumer Protection Act and the Construction (Design and Management) Regulations 1994 provide specific requirements for information to be provided.

2.7 Compatibility 512 331

All equipment should be selected and erected so that it will not be susceptible to interference from, or cause harmful effects to other equipment when used for the purpose for which it was intended, nor impair the supply during normal service, including switching operations (see Section 2.4 and 2.6 also). 512-05 331-01-01

This Regulation is of particular importance when considering supplies to information technology equipment. This includes computers, electronic office equipment, data transmission equipment and point of sale terminals. The power supply units of this type of equipment are particularly well known for harmonic generation problems.

Whenever electrical equipment is switched on or off, particularly where inductive loads are involved, high frequency voltage transients occur. This so-called mains-borne noise may cause malfunction in information technology equipment. The capacitance of the circuit cables and the filters incorporated in most information technology equipment attenuate this mains-borne noise. It is good practice to ensure that sources likely to give rise to significant noise (e.g. motors and thermostatically controlled equipment) are kept apart from sensitive equipment. Susceptible electronic equipment should be fed by separate circuits from the incoming supply point of the building. Additional filters (sometimes called 'power conditioners') may be used to reduce this transient noise on existing circuits, and surge protection devices can be installed to divert or absorb transients.

Large switch-on (inrush) current surges which occur with transformers, motors and mains rectifier circuits, can cause excessive short time voltage drop in the circuit conductors (dips). Inrush currents affect other circuits also and may require larger cable sizes, or the equipment to be on separate circuits. It should be borne in mind that information technology equipment itself can cause the same problem if switched on in large groups.

Where the maintenance of the supply voltage to especially sensitive equipment such as information technology equipment is considered of importance, the user may need a device such as a motor alternator or uninterruptible power supply (UPS).

When information technology equipment is planned functional earthing and a 'clean' (low-noise) earth must be considered. (See also Guidance Notes 5 & 7). 545 607-02-01

The protective conductors in a building are subject to transient voltages relative to the general mass of earth. These transients are termed 'earth noise' and are usually caused by load switching. They may be generated by the charging of an equipment frame via the stray capacitance from the mains circuit or mains-borne transients may be coupled into the earth conductor or frame from mains conductors.

As 'earth noise' can cause malfunction, manufacturers of large computer systems usually make specific recommendations for the provision of a 'clean' mains supply and a 'clean' earth. The equipment manufacturer's guidance must be taken for such installations.

A dedicated earthing conductor may be used for a computer system, provided that:

(i) all accessible exposed-conductive-parts of the computer system are earthed, the computer system being treated as an 'installation' where applicable 542-04 607-02

(ii) the main earthing terminal or bar of the computer system ('installation') is connected directly to the building main earthing terminal by a protective conductor 543-01

(iii) extraneous-conductive-parts within reach of the computer systems are earthed but not via the protective conductor referred to in (ii) above 413-02-03

Supplementary bonding between extraneous-conductive-parts and the accessible conductive parts of the computer system is not necessary.

Guidance Note 7 gives more information on Section 607 of BS 7671.

2.8 LV switchgear and controlgear assemblies — Forms of separation

BS EN 60439-1 : 1994 has replaced BS 5486 : Part 1 : 1990 which has been withdrawn. BS 5486 : Part 1 : 1990 remains applicable for use with other parts of BS EN 60439 (e.g. Parts 2, 3 and 4). This standard gives guidance on the forms of separation applicable to factory-built switchgear and controlgear assemblies (Motor Control Centres etc.), now known as type-tested and partially type-tested assemblies. These forms of separation provide protection against contact with live parts belonging to adjacent devices and protection from the probability of initiating arcing faults and the passage of foreign bodies between units of the assembly. The Standard also gives guidance on other requirements for protection against electric shock.

Definitions given in BS EN 60439-1 are:

> *Type-tested low voltage switchgear and controlgear assembly (TTA)*
>
> *A low voltage switchgear and controlgear assembly conforming to an established type or system without deviations likely to significantly influence the performance from the typical ASSEMBLY verified to be in accordance with this standard.*

For various reasons, for example transport or production, certain steps of assembly may be made in a place outside the factory of the manufacturer of the type-tested ASSEMBLY. Such an ASSEMBLY is considered as a type-tested ASSEMBLY provided the site assembly is carried out in accordance with the manufacturer's instructions which should be retained for future information. The instructions will include any requirements for submission of the ASSEMBLY to further applicable routine test.

> *Partially type-tested low voltage switchgear and controlgear assembly (PTTA)*
>
> *A low voltage switchgear and controlgear assembly, containing both type-tested and non-type-tested arrangements provided that the latter are derived (e.g. by calculation) from type-tested arrangements which have complied with the relevant tests.*

It should be noted that the forms of separation have no bearing on the overall ingress protection (IP rating or classification) for the external frame and enclosure of the switchgear assembly etc. This should be specified to the required IP rating (see Appendix B). For equipment for indoor use where there is no requirement for protection against ingress of water, the following IP references are preferred by BS EN 60439-1:

IP00, IP2X, IP3X, IP4X, IP5X.

Where some degree of protection against ingress of water is required, the following IP ratings are preferred:

TABLE 3
IP codes for protection against ingress of water

First Characteristic numeral Protection against contact and protection against ingress of solid foreign bodies	Second Characteristic numeral Protection against harmful ingress of water				
	1	2	3	4	5
2	IP21				
3	IP31	IP32			
4		IP42	IP43		
5			IP53	IP54	IP55
6				IP64	IP65

For equipment for outdoor use having no supplementary protection e.g. protective roof, the second characteristic numeral should be at least 3, depending on the installation conditions and exposure.

BS 5486 Part 1 : 1990 is directly equivalent to BS EN 60439-1 : 1994, but may have differences in constructional requirements from IEC 439-1: 1992. Designers and installers working on international projects should clarify specific requirements with the manufacturer.

Four forms of separation are indicated in the BS EN standard, but there is no specific detail given on how these forms are to be achieved. It is stated in the BS EN that the form of separation should be agreed between manufacturer and designer/user. It must be remembered that higher forms of separation specified will increase costs but will give better operational flexibility regarding safer working when connecting in additional circuits or carrying out maintenance. This 'trade off' must be carefully assessed.

The four forms given have basic definitions and applications, but Forms 2 to 4 can be further subdivided into more specific "Types" (applications) by discussion and agreement with manufacturers.

Form 1

This form provides for an enclosure to provide protection against direct contact with live parts, but does not provide any internal separation of switching, isolation or control items or terminations. These overall assemblies are often known as 'wardrobe' type with large front opening doors, usually with an integral door interlocked isolator. Operating the isolator interrupts all functions but allows the door to be opened to gain access to the assembly for installation or maintenance. Such assemblies normally have lower fault withstand and it may be inconvenient to shut down a whole plant or system for a simple maintenance or repair operation.

Fig 2: **Form 1 Construction**

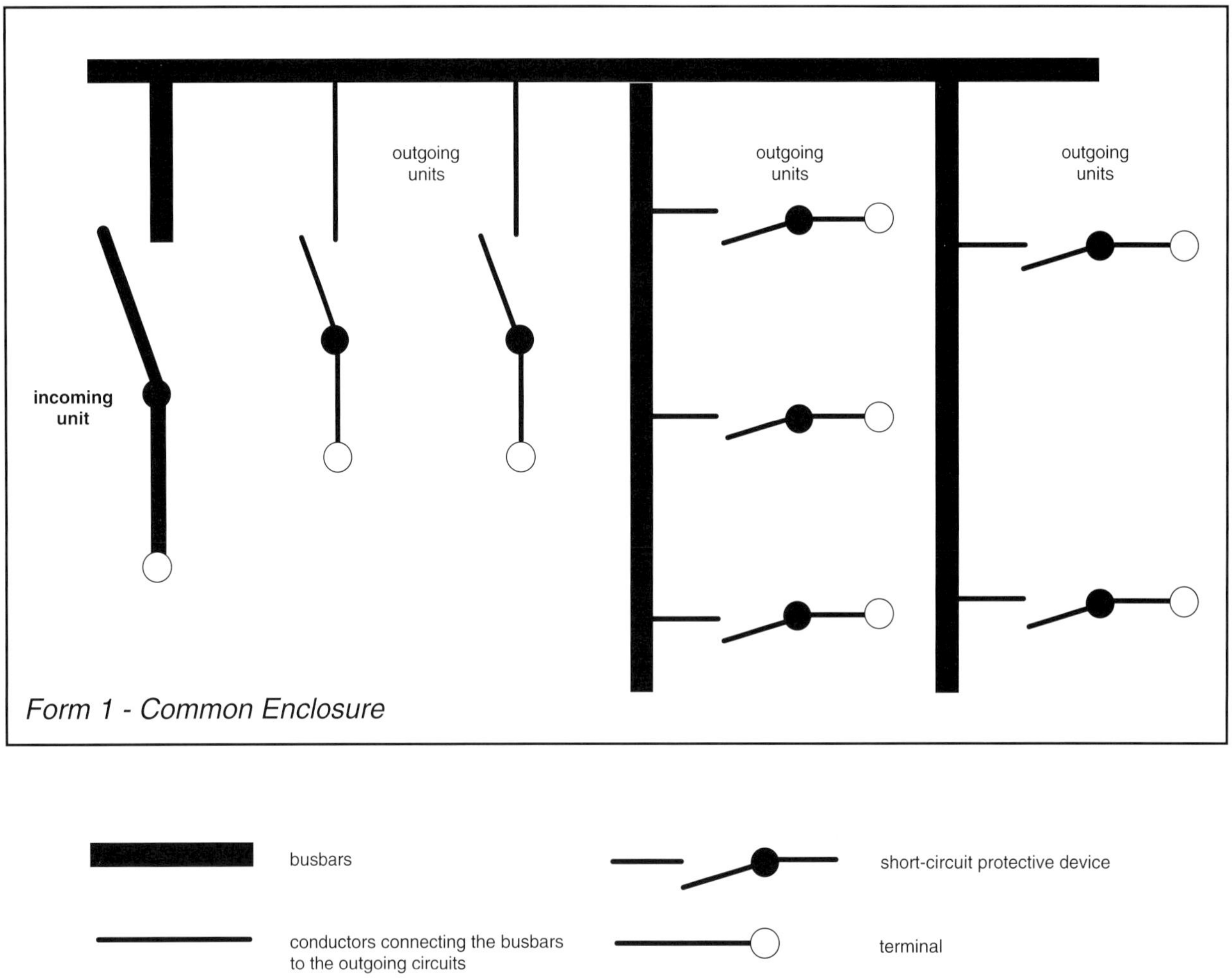

Form 2

The overall assembly enclosure provides protection against direct contact with live parts; separation is provided between the busbar assembly and switching, isolation, control items and terminations. There is very little advantage of this over Form 1, and the style is similar. Form 2 can be further subdivided into:

— Type 1, in which the busbars are separated by insulation of the bars

— Type 2, in which the busbars are separated by metallic or non-metallic rigid barriers.

Form 3

The enclosure provides protection against direct contact with live parts, and also separation is provided between the busbars and switching, isolation or control items, and between all these items. Outgoing terminals are not separated from each other, or perhaps from the busbars. Form 3 can be further subdivided into:

— Form 3a, in which outgoing terminals are not separated from the busbars.

— Form 3b, in which outgoing terminals are separated from the busbars.

Form 3b can be further subdivided into:

— Type 1, in which the busbars are insulated for separation.

— Type 2, in which busbar separation is by metallic or non-metallic rigid barriers or partitions.

Form 4

The enclosure provides protection against direct contact with live parts, and internal separation of the busbars from all switching, isolation and control items and outgoing terminations, and separation of all items and outgoing terminations from each other. This allows for access to any single item, such as a switch fuse or starter, and its outgoing terminations, to enable work to be carried out whilst the assembly remains operational. Protection is also provided against an arcing fault in one device affecting other items. This is the usual form specified for commercial and industrial switchgear and controlgear assemblies, but the designer has to consider whether due to the extra cost such requirements are necessary or justified. Form 4 can be further subdivided into seven types:

— Type 1, in which the busbars are separated by insulation coverings. Terminals for external conductors are in the same compartment as the associated item of switchgear etc., but cables may be glanded elsewhere.

— Type 2, in which the busbars are separated by metallic or non-metallic rigid barriers or partitions.

Terminals for external conductors are in the same compartment as the associated item of switchgear etc., but cables may be glanded elsewhere.

— Type 3, in which separation requirements are by metallic or non-metallic rigid barriers or partitions. Terminals for external conductors are in the same compartment as the associated item of switchgear etc., and each item has its own integral cable glanding facility.

— Type 4, in which the busbars are separated by insulated coverings. Terminals for external conductors are not in the same compartment as the associated item of switchgear etc., but in separate enclosed spaces, but cables may be glanded elsewhere.

— Type 5, in which busbars are separated by metallic or non-metallic rigid barriers or partitions. Terminals for external conductors are not in the same compartment as the associated item of switchgear etc., but in separate enclosed spaces, and terminals may be separated by insulated coverings. Cables may be glanded in common cabling chambers.

— Type 6, in which all separation requirements are by metallic or non-metallic rigid barriers or partitions. Terminals for external conductors are not in the same compartment as the associated item of switchgear etc., but in separate enclosed spaces, and cables are glanded in common cabling chambers.

— Type 7 in which all separation requirements are by metallic or non-metallic rigid barriers or partitions. Terminals for external conductors are not in the same compartment as the associated item of switchgear etc., but in separate enclosed spaces, and the termination for each item has its own integral glanding facility.

Fig 3: **Form 4 Construction (Type 7)**

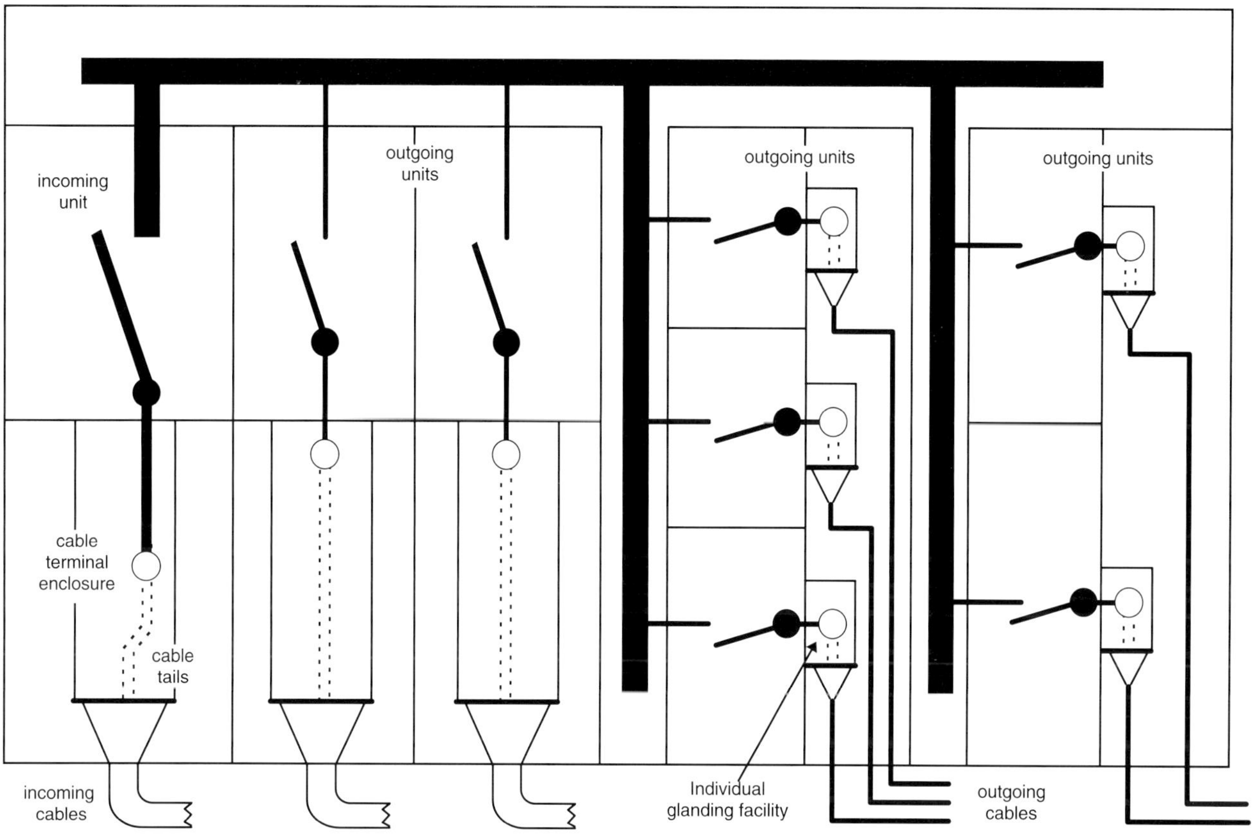

GENERAL

Assemblies are to be designed and constructed so as to be able to withstand the thermal and dynamic stresses resulting from fault currents up to their rated values. The designer must specify the prospective fault conditions at the point of installation.

Busbar systems for switchgear and controlgear should be adequately rated for the normal duty and maximum fault current level expected, and should be well supported and braced, as the electromechanical stresses under fault conditions can be severe. However thermal movement must be allowed for. It is usual for manufacturers to prove their busbar designs with a full short-circuit test by an independent test station up to the rated fault current level on a sample construction. The busbar arrangement can then be certified as a 'type-tested' busbar system.

Tests, as detailed in BS 5486 : Part 1, are applied to the complete assembly as relevant at manufacture, including continuity, insulation resistance and perhaps a flash test at high voltage. The designer and installer must be aware that there are statutory requirements under The Electricity at Work Regulations and The CDM Regulations etc. for the safe design, construction, operation and provision for maintenance of electrical equipment assemblies. Adequate access, working space and lighting must be provided where work is to be carried out on or near equipment, in order that persons may work safely.

It must also be realised that Regulation 8 of the Electricity at Work Regulations places an absolute requirement (one that must be met regardless of cost or any other consideration) on protective conductor connections to earth. "*A conductor shall be regarded as earthed when it is connected to the general mass of earth by conductors of sufficient strength and current carrying capability to discharge electrical energy to earth*". It may be questioned whether the termination of steel or aluminium wire armour (where used as a protective conductor) with glands into metal gland plates, which themselves may only be bolted to the switchgear or controlgear frame, is an adequate connection. Cable gland "earth tags" and supplementary connections to the equipment earth terminals may be necessary. In any case it should ensured that any paint or finish on the switchgear does not prevent good contact.

During site installation and commissioning, tests as required by BS 7671 Part 7 should be carried out on the complete assembly, plus any other specific tests advised by the manufacturer or required by the client, user or engineer. It is not usual to carry out a repeat of specialist tests such as a flash test at site, and in the event of a requirement the manufacturer's advice should be sought.

Section 3 — Protection Against Overcurrent and Electric Shock

3.1 Protective devices 531

The Regulations prescribe requirements for the following devices:

(i) overcurrent protective devices

(ii) residual current devices

(iii) insulation monitoring devices.

412

Protection against indirect contact which relies upon earthed equipotential bonding and automatic disconnection of the supply is intended to reduce the risk of shock by limiting the time and voltage magnitude in the event of a fault. Where it is not possible to achieve a sufficiently low fault loop impedance to operate an overcurrent device an RCD is usually used for protection from indirect contact. Supplementary equipotential bonding may also be provided to reduce the voltage magnitude. Low fault loop impedance however leads to high fault current and fast clearance times. 413-02 471-08 413-02-04

During a fault, voltages are present along the fault path. The protection must ensure that any such voltages are not likely to persist long enough to be dangerous. Access to such voltages may result in electric shock. The equipotential bonding will have voltages present along it if it forms part of the fault path. Supplementary equipotential bonding may be required in order to keep the magnitude of the voltages between simultaneously accessible exposed and/or extraneous-conductive-parts low. (See Part 6 of BS 7671 for specific requirements in special locations. Guidance Note 7 also addresses these requirements).

TABLE 4
Comparison of overcurrent protective devices

Semi-enclosed fuses	Cartridge fuses	Miniature circuit-breakers
Lowest capital costs	Higher capital costs	Highest capital cost
Low maintenance costs	Higher maintenance costs	No maintenance cost
Lowest fault current capacity	Highest fault current capacity	Intermediate fault current capacity
Requires a degree of skill to replace fusewire, fitting of wrong size/rating wire is not prevented	Relatively simple to replace cartridge. Size of cartridge relates to size of fuse carrier in consumer units for domestic installation. Easy to overwire.	Easy to reset. Unskilled persons able to reset. No replacement or refitting after operation, unless damaged by operation at high fault currents. Can operate on surges.

3.2 Overcurrent protective devices 533

Every circuit must be provided with a means of fault current and overload protection although some parts of a circuit — such as the flexible cord of a luminaire pendant — may be protected against overload by the nature of the load. The devices may be either fuses or circuit-breakers. For TN systems such overcurrent devices can also be used for indirect shock protection, provided that the earth fault loop impedance of the protected circuit is such that the chosen device will operate within the times specified in BS 7671. Regulation 413-02-19 also allows the use of the overcurrent protective device in a TT installation if the earth loop impedance is sufficiently low, and it can be certain that it will remain so. However, RCDs are preferred and usually specified. For TN systems Tables 41B and D of BS 7671 give maximum permitted earth fault loop impedances. The Table 41C gives maximum permitted impedances of circuit protective conductors corresponding to certain types of devices and current ratings of fuses and circuit-breakers. These

533-02
413-02-06
413-02-07
413-02-08
Table 41B
Table 41C
Table 41D
473-01-04

impedances provide an alternative method of providing earth fault protection. See Guidance Note 5 for further information. For other types of devices and current ratings the manufacturer should be consulted.

As indicated in Section 471, these limiting values of impedance are related to normal dry conditions. For the special locations dealt with in Part 6 of BS 7671 lower values may be required together with additional precautions. 471-01-01 471-08-01

The choice of protective device will depend on a number of factors, including overall installation and maintenance costs, and Table 4 gives a brief comparison between fuses and circuit breakers.

With respect to electric motors note should be taken of BS EN 60947-4-1. This recognises three types of co-ordination with corresponding levels of permissible damage to the starter. Co-ordination of Type 2 in BS EN 60947-4 (i.e. no damage, including permanent alteration of the characteristics of the overload relay, except light contact burning and a risk of contact welding) is required if the starter is to continue to provide overload protection complying with Regulation 433-02-01. Examination and maintenance of the devices is always necessary after a fault. 435-01-01 433-02-01

The selected fuse or circuit-breaker provides short-circuit protection for the motor circuit and the starter overload relay provides protection from overload. In motor circuits the short circuit protection and overload protection are often provided separately. Overloads relays must be selected for the motor duty, and for unusual duties — i.e. frequent starting/ stopping — the manufacturer should be consulted.

The overload relay on the starter is arranged to operate for values of current from just above full load to the overload limit of the motor, but it has a time delay such that it does not respond to either starting currents or fault currents. This delay provides discrimination with the characteristics of the associated fuse or circuit-breaker at the origin of the circuit on fault currents, and prevents operation due to surge currents at start up.

The starter overload relay can provide overload protection for the circuit in compliance with Regulation 433-02-01 on the basis that: 433-02-01

(i) its nameplate full load current rating or setting is taken as I_n, and

(ii) the motor full load current is taken as I_b, and

(iii) the ultimate tripping current of the overload relay is taken as I_2.

Where the overload relay has a range of settings then items (i) and (iii) should be based on the highest current setting, unless the setting cannot be changed without the use of a tool. (See Guidance Note 6 for further information).

3.3 Fuses 533

Fuses have a rated short-circuit capacity, see Table 5, and should be selected such that their rating is not exceeded by the prospective fault current at the point of installation, unless adequate back-up of protection is specified.

BS 88 Part 2 and Part 6 cartridge fuses have the same characteristics, as shown in Appendix 3 of BS 7671. 533-01-04 Appx 3

The size of a tinned copper single-wire element is specified in Table 53A for use where BS 3036 semi-enclosed (rewireable) fuses are selected and there are no manufacturer's instructions. Table 53A

BS 3036 fuses have a limited fault current-breaking capacity and also cannot be relied upon to operate within 4 hours at 1.45 times the nominal current of the fuse element. Correct protection can be obtained by modifying the normal condition $I_n \leq I_z$ such that the fuse rating does not exceed 1.45/2 = 0.725 times the rating of the circuit conductor. For this reason larger cables may need to be selected where overload protection is provided by semi-enclosed fuses than when it is provided by a cartridge fuse or fuses or circuit-breaker (mcb). (Appendix 4 of BS 7671 and Guidance Note 6 give further guidance). 433-02-03 Appx 4

Note should be taken of the warning regarding the possible replacement of a fuse link by one of a higher rating. 533-01-01

TABLE 5
Rated short-circuit capacities of fuses

Device type	Device designation	Rated short-circuit capacity kA
Semi-enclosed fuse to BS 3036 with category of duty	S1A S2A S4A	1 2 4
Cartridge fuse to BS 1361 type I type II		16.5 33.0
General purpose fuse to BS 88 Part 2.1 Part 6		50 at 415 V 16.5 at 240 V 80 at 415 V

3.4 Miniature circuit-breakers (mcbs) and circuit-breakers (cbs)

There is a wide range of breaker characteristics that have been classified according to their instantaneous trip performance and Table 4 gives some information on the applications of the various types available. These limits are the maximum allowed in miniature circuit breakers (mcbs) BS 3871 (now withdrawn) and circuit breakers (cbs) BS EN 60898 but it should be noted that manufacturers may provide closer limits. For such circuit-breakers, manufacturers' data for I_a may be applied to the formula given in Regulation 413-02-08. (I_a is the current causing automatic operation of the device within the stated time.)

Appx 3

413-02-08

TABLE 6

Circuit-breakers overcurrent protection

Type	Multiple of rated current I_n below which it will not trip within 100 ms	Multiple of rated current I_n above which it will definitely trip within 100 ms	Typical application
1* B	2.7X 3X	4X 5X	Circuits not subject to inrush currents/ switching surges
2* C	4X 5X	7X 10X	Circuits where some inrush current may occur. For general purpose use on fluorescent lighting circuits, small motors etc
C 3*	5X 7X	10X 10X	Circuits where high inrush currents are likely, e.g. motors, large lighting loads, large air conditioning units
D 4*	10X 10X	20X 50X	Circuits where inrush currents are particularly severe, e.g. welding machines, x-ray machines.

Notes:

1. Some non-linear resistive loads, such as large tungsten filament lighting installations, may give rise to high inrush currents.
2. * Due to the withdrawal of BS 3871, Types 1 to 4 mcbs are now obsolete, but are still to be found in use.
3. Type D and Type 4 cbs are special purpose breakers which require low EFLI and should not be used without due consideration, particularly where this device is to effect protection against indirect contact. Data should be obtained from the manufacturer.

All mcbs and cbs have a maximum fault current breaking capacity and care is needed in selection to ensure that this is not exceeded in service. BS 3871 identified this capacity with an 'M' rating, however in BS EN 60898 the 'M' category ratings for breaking capacities are replaced by a figure in a rectangle (I_{cn}) on the device (see Table 7 below). The manufacturer must declare the breaking capacity of the devices at specified power factors of test circuit. Higher fault current capacities up to 25 kA are recognised for BS EN 60898 devices. Rated values for both standards are given in the following table:

TABLE 7 **Comparison of rated values of miniature circuit-breakers to BS 3871 and BSEN 60898 circuit-breakers breaking capacity**

BS 3871		**BS EN 60898**			
Marked on device as 'M' number		I_{cn} (rated capacity) (Marked on device in rectangle)		k	I_{cs} (service capacity)
M1	1000 A				
M1.5	1500 A	1500	1500 A	1	1500 A
M3	3000 A	3000	3000 A	1	3000 A
M4.5	4500 A	4500	4500 A	1	4500 A
M6	6000 A	6000	6000 A	1	6000 A
M9	9000 A				
		10000	10000 A	0.75	7500 A
		15000	15000 A	0.5	7500 A
		20000	20000 A	0.5	10000 A
		25000	25000 A	0.5	12500 A

(k — ratio between service short-circuit capacity and rated short-circuit capacity, BS EN 60898).

Two short-circuit capacities I_{cn} and I_{cs} are quoted for circuit-breakers.

I_{cn} the rated short-circuit capacity. (Ultimate short-circuit breaking capacity).

I_{cs} the service short-circuit capacity. This is the maximum level of fault current operation after which further service is assumed without loss of performance.

For an assigned I_{cn} the I_{cs} value will not be less than the value tabulated in Table 7. The I_{cn} of the circuit-breaker must always exceed the prospective short-circuit current at the point of installation, except where backup protection as specified by the manufacturer is applied.

(Moulded case circuit breakers to BS EN 60947-2 do not have defined characteristics as do circuit breakers to BS 3871 or BS EN 60898 and the manufacturers data must be used.)

3.5 Insulation monitoring devices 413

Devices which monitor and indicate the condition of the insulation must be installed in IT systems to indicate a first fault to earth. Such systems may also be used where high degrees of reliability of supply are necessary, as in supplies for safety services. 413-02-24 561-01-03

The first fault on such a system would then signal the need for remedial action allowing time to carry out the procedure, before a possible second fault arose.

Such a system calls for a special knowledge and is outside the scope of this Guidance Note. An IT system is not allowed for low voltage public supply in the UK. 120-02-01 Their use is generally confined to industrial control systems, or hospital operating theatres and treatment rooms.

3.6 Residual current operated devices (RCDs) 531

A core balance transformer assembly is used to detect the existence of an earth fault. This recognises the out-of-balance current between the circuit live conductors (the residual current) that an earth leakage fault produces. A current is then induced which is used to operate the tripping mechanism of a contact system to interrupt the circuit.

Particularly in larger circuits residual current protection may be provided by an assembly of detectors and relays. Where a power source is needed in order to trip the circuit breaker concerned steps must be taken to ensure that the tripping supply is available at all times

There are now several terms in use with RCD products, which are described below.

RCD (Residual current device) is the generic term for all products which use the principle of detecting earth leakage fault current by measuring the difference in current magnitude flowing in different supply conductors.

RCCBs (Residual current operated circuit-breakers) are circuit-breakers which incorporate the contact system within the same product.

SRCDs (Socket-outlets incorporating RCDs) are RCDs incorporated into a socket-outlet.

PRCDs (Portable residual current devices) are portable RCDs which can be plugged into a socket-outlet and which can supply equipment by various means e.g. socket-outlet, in line device, within the plug etc.

RCBOs (RCCBs with integral overcurrent protection) are RCCBs which also incorporate protection against overcurrent within the same product. RCBOs were previously known as 'combined mcb/RCCBs' (or sometimes 'combined mcb/RCDs'.

SRCBOs (Socket-outlets incorporating RCBOs) are RCBOs — RCCBs with integral overcurrent protection — connected to fixed socket-outlets.

RCCBs currently meet the requirements of BS 4293 which is being replaced by European standard BS EN 61008-1. RCBOs are designed to the requirements of BS EN 60898 (or BS 3871) and BS 4293, but these are being replaced by the European standard BS EN 61009-1. SRCDs should comply with BS 7288 and PRCDs with BS 7071. Residual current devices to these BS EN standards are manufactured to the degree of protection IP20. (See

Appendix B). Any other degree of protection will be marked on the device.

International and European standards are being prepared for SRCDs and PRCDs. In addition, a European standard for SRCBOs is being prepared.

RCDs for load currents below 100 A usually include the transformer and contact system within the same enclosure. Devices for load currents greater than 100 A usually comprise a transformer assembly with a detector and a separate shunt trip circuit-breaker unit or contactor, mounted together.

A wide choice of residual operating current ratings ($I_{\Delta n}$) is available, typical values being between 10 mA and 2 A. Some RCDs are manufactured so that one of several operating currents may be selected. These types of RCD must not be installed where they would be accessible to unauthorised persons, or a type in 531-02-10
which a deliberate act with a key or tool is required to adjust the setting must be installed. Single-phase or multi-phase devices, with or without integral overcurrent protection, are available. RCCBs however are not adjustable.

Where RCDs have rated residual operating currents of 30 mA or less there is a choice between types of devices. Some are entirely electromechanical in operation and others use solid state devices. These are classified in BS EN 61008 and BS EN 61009 as RCCBs or RCBOs 'functionally independent of line voltage' or 'functionally dependent on line voltage' respectively. Types functionally dependent on line voltage are required by the standard to be identified. The electromechanical types operate simply on the power being fed to the fault whereas the solid state types require a power supply for operation. BS 7671 states that RCDs requiring an auxiliary supply and 531-02-06
which do not operate automatically in the case of failure of that source shall be used only if:

(i) protection against indirect contact is maintained by other means, or

(ii) the device is in an installation operated and maintained by an instructed person or skilled person (because such persons should be aware of the risk).

A test device is incorporated to allow the operation of the RCD to be checked. Operation of this device creates an out-of-balance condition within the device which establishes the integrity of the electrical and mechanical elements of the tripping device only. It should be noted that the test device does not provide a means of checking the continuity of any part of the earth path nor does it check the minimum operating current or operating time of the RCD.

The introduction of BS EN 61008 and BS EN 61009 extends the classification of residual current devices beyond that of the old UK standard BS 4293, and include classification of such attributes as time delay facilities and operating characteristics for currents with d.c. components. RCDs are now categorised into three types 'A' and 'AC', each of which is available in the 'General' type — without any operating time delay — and the 'S' (Selective) type — which has a built in operating time delay to provide discrimination of RCDs when connected in series (see below). An RCD used for supplementary protection against direct contact must not have a time delay in operation and 'General' types must be used. The 'Selective' type is used to provide discrimination. In a TN system the time delay must not exceed the requirements of Regulation 413-02. 413-02

There also exists a further type 'Class B' which is manufactured to IEC 755, but has no BS or BS EN equivalent standard. This is a specialist type and is for operation on pure d.c., impulse d.c. and a.c.

A	Tripping is ensured for residual sinusoidal alternating currents and pulsating direct currents	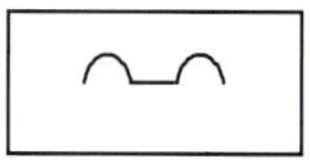
AC	Tripping is ensured for residual sinusoidal alternating currents	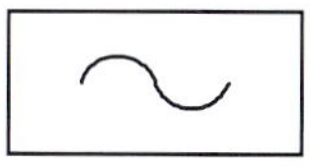

Type A devices are important due to the wide range of electronic equipment available now which may produce a leakage current with a pulsating d.c. component and harmonic currents. These RCDs may utilise electronic circuitry or low remanence core material to more closely match the predetermined current time trip characteristics allowed by the standards to achieve the most suitable performance characteristics.

Where, in order to minimise the risk of danger and inconvenience to the user, discrimination is required between residual current devices operating in series, the device operating characteristics must provide the required discrimination. A time delay should be provided in the 'upstream' device by the use of a Type S device. Type S devices are identified by the symbol S marked on the device 531-02-09

Tables 8 and 9 below (extracted from BS EN 61008 and BS EN 61009) provide comparative data for the devices.

TABLE 8
Standard values of break time and non-actuating time for Type AC RCCBs (extract from BS EN 61008)

Type	I_n	$I_{\Delta n}$	Standard values of break time(s) and non-actuating time(s) at a residual current (I_Δ) equal to:				
	A	A	$I_{\Delta n}$	$2I_{\Delta n}$	$5I_{\Delta n}$	500 A	
General	Any Value	Any Value	0.3	0.15	0.04	0.04	Maximum break times
S	≥ 25	> 0.030	0.5	0.2	0.15	0.15	Maximum break times
			0.13	0.06	0.05	0.04	Minimum non-actuating times

TABLE 9

Standard values of break time and non-actuating time for operating under residual conditions for Type AC RCBOs (extract from BS EN 61009)

Type	I_n	$I_{\Delta n}$	Standard values of break time(s) and non-actuating time(s) at a residual current (I_Δ) equal to:				
	A	A	$I_{\Delta n}$	$2I_{\Delta n}$	$5I_{\Delta n}$	$I_{\Delta t}$	
General	Any Value	Any Value	0.3	0.15	0.04	0.04	Maximum break times
S	≥ 25	> 0.030	0.5	0.2	0.15	0.15	Maximum break times
			0.13	0.06	0.05	0.04	Minimum non-actuating times

Definitions:

I_n = rated current of the device.

$I_{\Delta n}$ = rated residual operating current of the device.

$I_{\Delta t}$ = value of current ensuring residual current sensor does not operate before the overload sensor.

The purpose of an RCD in any installation is to provide protection against earth leakage currents.

RCDs with a suitable differential current rating and operating time are therefore suitable for:

— protection against indirect contact 413-02-04

— some supplementary protection against direct contact 413-06

— some protection against the risk of fire caused by earth leakage currents. 605-10-01

RCDs are intended to operate to provide protection against phase-to-earth and neutral-to-earth faults but do not operate in the event of phase-to-phase or phase-to-neutral faults.

An RCD will only restrict the time during which a fault current flows. It cannot restrict the magnitude of the fault current which depends solely on the circuit conditions. 412-06

For an installation which is part of a TN system the use of an RCD may be from choice or because the earth fault loop impedance is too high for 413-02-04 413-02-07 overcurrent protective devices to operate within the maximum permitted disconnection times. In the latter case, a further option is to use supplementary equipotential bonding so that there will be no dangerous potential between simultaneously accessible exposed- and extraneous-conductive-parts.

When an RCD is used the product of the rated residual operating current ($I_{\Delta n}$), in amperes, and the earth fault loop impedance (Z_s), in ohms, must not exceed 50 V. (This requirement is reduced further for some special locations in Part 6 of BS 7671.) This does not mean, however that the voltage to earth will not exceed 50 V or such other prescribed voltage; its purpose is to ensure that when the voltage does exceed the prescribed voltage the RCD will operate. 413-02-16

Fig 4: **Installing RCDs in a TT installation**

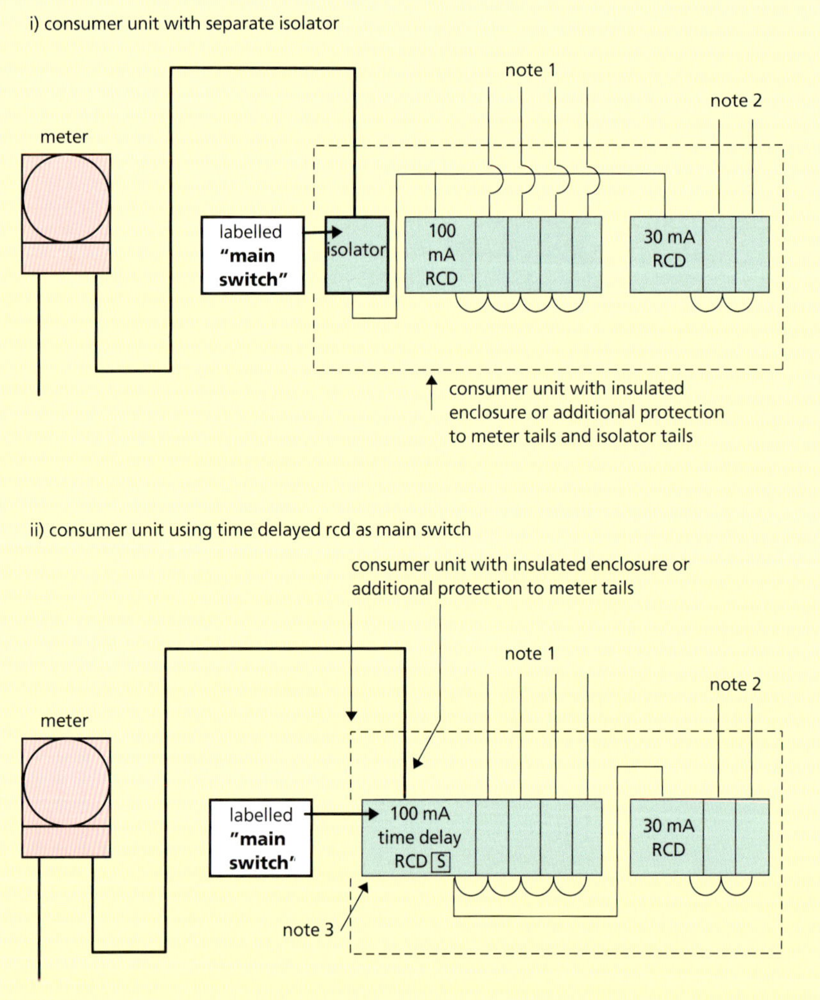

note 1 circuits to lights, cooker, water heater
2 circuits to portable equipment outdoors and socket outlets that may reasonably supply portable equipment outdoors (ground floor socket outlets)
3 time delay or S-type double pole RCD to BS EN 61008. This RCD must be clearly identified (labelled) as the isolating switch for the installation

BS 7671 requires that an RCD be used as supplementary protection against direct contact for any socket-outlet rated at 32 Amps or less that may reasonably be expected to supply portable equipment for use outdoors in a domestic, commercial or industrial installation. As indicated in the Regulations, the RCD has then to have a rated residual operating current not exceeding 30 mA and an operating time not exceeding 40 ms at a residual current of 150 mA. This requirement is met by general type devices complying with BS EN 61008 or BS EN 61009 (or BS 4293). 471-16-01 412-06-02

This RCD must be part of the fixed wiring of the installation and in domestic or residential installations can either be installed at the consumer unit as a separate RCD on all socket-outlet circuits (i.e. a split board) or a combined RCD/mcb device (RCBO). An alternative is to specifically provide a socket-outlet with integral RCD (SRCD) to BS 7288 at each location where a supply may be taken to be used outdoors in domestic, commercial or industrial installations (e.g. in a kitchen, garage, front porch, conservatory, workshop yard etc.). Obviously, socket-outlets on the first floor of a house or in flats above the ground floor will not normally be suitable for use to supply equipment outdoors and so would be unlikely to need RCD protection for this purpose.

It is not a good practice to provide a single RCD on the supply to a complete consumer unit, which may have lighting circuits etc., as operation will interrupt the whole building supply. Regulation 314-01-01(i) requires installation circuit arrangements to be such that danger is avoided as far as reasonably practicable and inconvenience minimised in the case of a fault. It can be argued that protection of an installation by an RCD is not related to the 'circuit' definition of Part 2 of BS 7671, and Regulation 531-04-01 makes specific reference to a single RCD for a TT installation. However, the decision to provide a single RCD, to use a split consumer unit or to use RCBOs must be taken on consideration of all the relevant factors, including convenience and maintenance, and not just on initial costs. Elderly or infirm persons may not be able to gain access to, or reset, an RCD. On a TT installation, where overall RCD protection is required, it is 314-01-01 531-04-01 advisable to install a split consumer unit with a 30 mA general RCD on the socket-outlet circuits and a 100 mA time delayed S type on the other circuits. See Figure 4. A single 30 mA RCD should never be installed at the origin of the installation.

For some special locations (see Part 6 of BS 7671), eg Zones AB of swimming pools, if the protective measure against indirect contact is automatic disconnection of supply, the Regulations call for the use of an RCD having a rated residual operating current not exceeding 30 mA. Also, in an agricultural 605-10-01 installation an RCD with a residual operating current not exceeding 500 mA is required to be installed to provide some protection against fire caused by earth leakage currents. This device would not provide supplementary protection for direct contact electric shock.

BS 7671 recognises, in a TT system, that either RCDs or 413-02-17 overcurrent protective devices may be used to provide 413-02-18 protection against indirect contact. In many such 413-02-19 installations it will not be practicable to attain 413-02-20 sufficiently low values of earth electrode resistance to 471-08-06 use overcurrent protective devices. In any event all socket-outlet circuits in those installations have to be protected by an RCD. There is no specific limitation placed on the rated residual operating current, the sole criterion being that there should be compliance with BS 7671. Where the socket-outlet circuit is reasonably expected to supply portable equipment for use outdoors or the circuit installation is part of a special (Part 6) location, the foregoing guidance applies, and an RCD with a 30 mA maximum residual operating current must be provided.

From the foregoing it is seen that it is possible to categorise an RCD into one of two groups according to its fault current operating characteristics:

(i) an RCD having a rated residual operating current greater than 30 mA. This group is intended solely to give protection against indirect contact or fire.

100 mA, 300 mA and 500 mA RCDs are 605-10-01 sometimes said to protect against fire in the event of earth leakage. It should be noted that

these units do not offer protection from direct contact as afforded by 30 mA devices.

(ii) RCD's having a rated residual operating current of 30 mA or less and operating time not exceeding 40 ms at a residual current of 150 mA as provided by general type RCCBs to BS EN 61008 or RCBOs to BS EN 61009. This group is generally referred to as 'high sensitivity'. In addition to giving protection against indirect contact, they are recognised by BS 7671 as giving supplementary protection against **direct** contact. That is, protection of persons who come into simultaneous contact with a live part and earth. 412-06-02

When an RCD is used to give supplementary protection against direct contact, it is essential that a basic measure against direct contact is also used. The use of an RCD in a circuit normally expected to have a protective conductor, is not considered sufficient for protection for that circuit against indirect contact where there is no such protective conductor, not withstanding if the rated residual operating current of the RCD does not exceed 30 mA. 412-06-01 531-02-05

As well as providing protection against indirect and supplementary protection against direct contact, with a suitable differential current rating, RCDs may also provide some protection against fire risk, as stated above. The level of protection is related to the sensitivity of the device. For this purpose an RCD should be chosen with the lowest suitable rated residual operating current. A lower operating current would give a greater degree of protection, but it may also result in unwanted tripping and the connection of a further load at a later date may have an exacerbating effect, due to increase leakage.

In a circuit which is intended to have a protective conductor but does not, it is not sufficient to attempt to overcome the deficiency of a protective conductor by providing a RCD. 531-02-05

The installation designer will often not know the sum total of the earth leakage currents occasioned by the loads. Neither will it always be known what equipment is going to be used, nor (if a number of circuits are to be protected by one RCD) how many of 531-02-04 607 those equipments would be energised at any one time.

Knowing the use to which the installation will be put, the designer must deduce the likely total leakage current in the protected circuit or make an assessment and state this in the design. In cases of difficulty, circuits may be sub-divided to reduce the leakage and the effect of unwanted tripping. 531-02-04

The total leakage of the various items of equipment supplied by the RCD concerned should be such that any earth leakage current expected in normal service will not cause unwanted operation of the device. 531-02-04
Regulation 607-02-03 which refers to equipment with high earth leakage current requires that the total earth leakage current from stationary equipment installed in an installation with high earth leakage currents should not exceed 25% of the nominal tripping current of the RCD $I_{\Delta n}$. 607-02-03

British Standards with safety requirements for electrical equipment generally include limiting values of leakage current. Limits apply when cold and also at operating temperature. For instance, BS EN 60335-1 and BS 3456 Part 101 (which covers the general requirements for safety of household and similar electrical appliances) and other British Standards prescribe the limits shown in Appendix J.

Unwanted tripping (sometimes wrongly called nuisance tripping) of RCDs occurs when a leakage current causes unnecessary operation of the RCD. Such tripping may occur on heating elements, cooking appliances etc., which may have elements which absorb a small amount of moisture through imperfect element end seals when cold. When energised, this moisture provides a conductive path for increased leakage and could operate an RCD. The moisture dries out as the element heats up. Although it is not precluded in BS 7671, it is not a requirement to use an RCD on such circuits if other satisfactory earthing means of earthing are available. Providing an RCD with a higher residual operating current may solve the problem, but the requirements of the Regulations would still have to be met. For example the requirement to provide sensitive RCD protection for sockets likely to be used to supply portable/hand held equipment for use outside.

Further information on earth leakage protection and application of RCDs is provided in Guidance Note 5. Details of the testing of RCDs are given in Guidance Note 3.

3.7 Earthed equipotential bonding and automatic disconnection of supply 413

The tables of limiting impedances given in Chapter 41 of BS 7671 are for circuits having a nominal voltage to earth (U_o) of 230 V rms a.c. For other values of U_o the tabulated earth fault loop impedance values should be multiplied by U_o/230. 413-02

In Appendix 3 of BS 7671, curves have been drawn to represent the slowest mcb instantaneous operating times allowed by the standard and median times for fuses. To assist the designer, a set of time/current values for specific operating times has been agreed for each device and is reproduced at the right hand side of each set of curves. Appx 3

Regulations 413-02-09 and 413-02-13 of BS 7671 specify maximum disconnection times for circuits. 413-02
Regulations 413-02-10, 413-02-11 and 413-02-14 provide maximum earth fault loop impedances (Z_s) that will result in protective devices operating within the required disconnection times (of Regulations 413-02-09 and 413-02-13). 413-02-14

The maximum earth loop impedance for a protective device is given by: 413-02-08

$$Z_s = \frac{U_o}{I_a}$$

Where:

U_o is the nominal voltage to earth (the open circuit voltage at the distribution transformer is used in Appendix 3 as U_{oc}).

I_a is the current causing operation of the protective device within the specified time.

For the purposes of Regulations 413-02-10 and 413-02-13 (Tables 41B1, 41B2, 41D, 604B1, 604B2, 605B1 and 605B2) the open circuit voltage U_{oc} has been presumed to be 240 V for nominal supply 413-02 voltage U_o of 230 V. The note in Appendix 3 of BS 7671 explains that this allows for the open circuit voltage at the remote distribution transformer.

The tabulated values in BS 7671 are generally applicable for nominal 230 V supplies within the statutory limits from regional electricity companies. For other supplies the designer will need to determine open circuit voltages and calculate Z_s accordingly.

Where it is not possible to obtain a sufficiently low earth fault loop impedance, there are other options for an installation which is part of a TN system. Either apply local supplementary equipotential bonding or use a residual current device. 413-02-04

Where supplementary equipotential bonding is to be installed it is necessary to connect together the exposed-conductive-parts of equipment in the circuits concerned including the earthing contacts of socket-outlets and extraneous-conductive-parts. Supplementary bonding conductors must be selected to comply with the minimum size requirements of Regulation 547-03, and the resistance (R) of the supplementary bonding conductor between simultaneously accessible exposed-conductive-parts and extraneous-conductive-parts must fulfil the following condition: 413-02-27 413-02-28

$$R \leq \frac{50}{I_a}$$

Where:

I_a is the operating current of the protective device:

(i) for a residual current device, the rated residual operating current $I_{\Delta n}$

(ii) for an overcurrent device, the minimum current which disconnects the circuit within 5 s.

However, if local supplementary bonding is provided to limit shock voltage magnitude there is still a requirement to disconnect the supply to the circuit for protection against thermal effects (see Appendix D). The circuit must be designed such that cables and equipment will not be damaged by the thermal effects of the fault current, whatever the disconnection time, as BS 7671 does not limit the disconnection time in this instance. (See Guidance Note 5 for further information.)

The application of RCDs is discussed in Section 3.6.

Table 41C provides maximum values for the impedance of a final circuit protective conductor (R_2) related to the rating of the final circuit protective device. This was known in the 15th Edition of the IEE Regulations as the 'Alternative Method' but has never been widely used. The protective conductor impedance is measured from the outlet to the main earth terminal (or point of connection of additional main equipotential bonding if installed to comply with Regulation 413-02-13(ii)). Regulation 413-02-12 allows circuits designed to this Regulation to have a disconnection time up to 5 seconds, including socket-outlet circuits. This is because the selection of such maximum protective conductor impedances will limit the duration and magnitude of voltages to earth under fault conditions to a level considered to be below normally dangerous levels. The calculation of the impedances given in Table 41C is beyond the scope of this Guidance Note.

It is worth noting the provisions of Regulation 543-02-02, and specially that steel wire armour, metal conduit and trunking etc. properly installed and maintained, are suitable as protective conductors where they can fulfil the shock, thermal protection requirements, and minimum size of Section 543 of BS 7671, and separate copper protective conductors are NOT always required. However, for functional reasons a functional earth connection may need a copper conductor. The manufacturer of the equipment should be consulted. 543-02-02

Main equipotential bonding is required to extraneous-conductive-parts including main services and metallic structural parts as given in Regulation 413-02-02. Conductors should be selected and installed in accordance with the requirements of Regulation 547-02-01 of BS 7671. Table 10 below gives guidance on required conductor sizes. 413-02-02 547-02-01

Supplementary equipotential bonding is installed as appropriate in accordance with the requirements of Section 547 of BS 7671 also. Table 11 below gives guidance on conductor sizes.

Protective multiple earthing (PME) is now widely used for low voltage supplies (a TN-C-S system), and has specific requirements for earthing and bonding of installation. These are discussed in more detail in Guidance Note 5.

It has been questioned whether flush metal accessory boxes are to be considered as "Exposed-conductive-parts" (as defined in Part 2 of BS 7671) and consequently whether they require to be connected specifically to the circuit protective conductor. For a time it has been accepted that tight metal to metal contact via a plate screw from the accessory earth strap (usually routed to the accessory fixing screw hole) to a fixed lug in the metal box was sufficient. The box was inaccessible in normal use and when the accessory was taken off the box the box effectively became a piece of isolated metalwork, and usually the supply to the circuit is disconnected. It has however recently been agreed by the Wiring Regulations National Committee that a protective conductor connection between the accessory and its box should be made, although this is not a Regulatory requirement.

TABLE 10

Main earthing and main equipotential bonding conductor sizes (copper equivalent) for TN-S and TN-C-S supplies

Phase conductor or neutral conductor of PME supplies	mm^2	4	6	10	16	25	35	50	70
Earthing conductor not buried or buried protected against corrosion and mechanical damage see notes	mm^2	4	6	10	16	16	16	25	35
Main equipotential bonding conductor see notes	mm^2	6	6	6	10	10	10	16	16
Main equipotential bonding conductor for PME supplies (TN-C-S)	mm^2	10	10	10	10	10	10	16	25

Notes:

1. Protective conductors (including earthing and bonding conductors) of 10 mm^2 cross-sectional area or less shall be copper.

2. Regional electricity companies may require a minimum size of earthing conductor at the origin of the supply of 16 mm^2 copper or greater for TN-S and TN-C-S supplies.

3. Buried earthing conductors must be at least:

   25 mm^2 copper if not protected against mechanical danger or corrosion
   50 mm^2 steel if not protection against mechanical damage or corrosion
   16 mm^2 copper if not protection against mechanical damage but protection against corrosion
   16 mm^2 coated steel if not protected against mechanical damage but protected against corrosion.

4. Regional electricity companies should be consulted when in doubt.

TABLE 11
Supplementary bonding conductors

Size of circuit protective conductor	Minimum cross-sectional area of supplementary bonding					
	Exposed-conductive-part to extraneous-conductive-part		Exposed-conductive-part to exposed-conductive-part		Extraneous-conductive-part to extraneous-conductive-part (1)	
	mechanically protected	not mechanically protected	mechanically protected	not mechanically protected	mechanically protected	not mechanically protected
mm^2	mm^2 1	mm^2 2	mm^2 3	mm^2 4	mm^2 5	mm^2 6
1.0	1.0	4.0	1.0	4.0	2.5	4.0
1.5	1.0	4.0	1.5	4.0	2.5	4.0
2.5	1.5	4.0	2.5	4.0	2.5	4.0
4.0	2.5	4.0	4.0	4.0	2.5	4.0
6.0	4.0	4.0	6.0	6.0	2.5	4.0
10.0	6.0	6.0	10.0	10.0	2.5	4.0
16.0	10.0	10.0	10.0	16.0	2.5	4.0

Notes:

1. If one of the extraneous-conductive-parts is connected to an exposed-conductive-part, the bond must be no smaller than that required for bonds between exposed-conductive-parts and extraneous-conductive-parts — column 1 or 2.

SECTION 4 — External Influences

4.1 External influences

The effect of environmental conditions and general characteristics around various parts of the installation must be assessed to enable suitable electrical equipment to be specified. 300-01

A list of external influences relating to Chapter 32 of BS 7671 is given in Appendix 5. These external influences are also identified in Section 522 of BS 7671. Chapter 52 of BS 7671 applies to the selection and erection of wiring systems, but the classification of external influences is applicable and relevant to all types of electrical equipment. Chap 32 Appx 5 522

If every part of the wiring system complies with the IP degree of protection for the worst circumstances expected in the particular location, and correct measures are taken during installation and maintenance, the selection and erection requirements of BS 7671 with respect to external influences will generally be satisfied. Due consideration must be given to the installation of equipment, including fixings, and manufacturers' installation instructions complied with, or the declared IP rating of the equipment may not be achieved. Good workmanship is, as always, required. 130-01-01

All electrical equipment selected must be suitable for its location of use and method of installation, and should not be modified on site unless it is designed and manufactured to allow this. Manufacturers' recommendations and instructions should be followed. BS 7540 : 1994 'Guide to use of cables with a rated voltage not exceeding 450/750 V' gives general cable selection guidance. 512-06-01

Equipment must also be protected against damage and ingress of foreign bodies during construction work. This damage and ingress may be more severe than operational conditions.

References are made in the following sections to certain aspects covered in Section 522. The coding used in Appendix 5 of BS 7671 is given in parenthesis.

4.2 Ambient temperature (AA) 522

Local ambient temperature means the temperature of the air or other medium around the equipment. There can be a wide range between the highest and lowest temperatures around equipment, depending on time of year, artificial heating etc. A design must be based on the most extreme ambient envisaged. In the UK external ambient temperatures between -10 °C and +35 °C commonly occur, and occasionally even lower or higher temperatures than these. 522-01

The cable current-carrying capacity tables in Appendix 4 of BS 7671 are based on an ambient temperature of 30 °C. Where this temperature differs, correction factors are to be used. Appendix 4 Clause 2.1 and Tables 4C1 and 4C2 should be consulted. Appx 4 Table 4C1 Table 4C2

Although the temperature in most homes in the UK would not usually be over 30 °C, close to a heating appliance it may be higher. BS 1363 accessories and cable with an operating temperature rating of 70 °C or lower are not recommended for installation in areas of higher ambient temperature such as airing cupboards, and in particular must not be covered with clothes or other items. Heat resisting cables appropriate to the ambient and equipment temperatures must be utilised. 522-01

Cables in heated floors or other parts must be suitable for the highest temperature reached.

Cables and equipment must be suitable for the lowest likely temperature. In particular, general purpose pvc cables should not be installed in cold places. Attention is drawn to the fact that, as the temperature decreases, pvc compounds become increasingly stiff and brittle, with the result that, if the cable is bent quickly to a small radius or is struck at temperatures in the region of 0 °C or lower, there is a risk of shattering the pvc components.

To avoid the risk of damage during handling, it is desirable that pvc-insulated or sheathed cables should be installed only when both the cable and the ambient temperature are above 0 °C and have been so for the previous 24 hours, or where special precautions have been taken to maintain the cables above this temperature. BS 7540 : 1994 advises a minimum ambient temperature of 5 °C for some types of PVC insulated and sheathed cables.

4.3 Solar and other radiation (AN) 522

Where cables are installed in situations subject to solar or other radiant heat sources, they should be resistant to damage from that source or effectively protected from damage. Cables may need to be derated where there are heat sources or heat is absorbed by solar gain. 522-02 522-11

To protect cables from the adverse effects of Ultra Violet (uv) radiation, the sheaths often contain carbon black or alternative uv stabilisers. Physically shielding the cables from direct solar radiation may also be necessary. In situations where high ambient temperatures are present (e.g. hot countries) cables will need to be derated. Shielding may be necessary, to protect from both direct heat and heat absorbed from solar radiation but cables should be freely ventilated. Cable manufacturers will provide guidance on the use of cables outdoors.

Where weather resistance is required the type of cable sheath to give protection should be carefully considered. Cables sheathed with pvc to BS 6746 or synthetic rubber to BS 6899 can give adequate performance but natural rubbers do not perform well. Cable sheathing materials should conform to BS 7655. Protection from mechanical damage may also be required. Additional advice is given in BS 7540.

The exposure of pvc insulation or sheathing to high temperatures for any length of time will lead to softening. If this temperature is maintained for long periods, the pvc can decompose and give off corrosive products which may attack the conductors and other metalwork. Softening of the pvc may also allow the conductor to move through the insulation if any mechanical force is applied (e.g. cables hanging vertically). Decentralisation of conductors may also take place where cables are overheated when bunched inside trunking.

4.4 The IP and IK classifications 522

The designer must assess each area of external influence and then ensure the required classification is selected and applied correctly. 512-06

The IP classification standard BS EN 60529 : 1991 identifies degrees of protection against the ingress of solids and water. It has replaced old descriptions such as 'ordinary', 'drip-proof', 'splash-proof' and 'water-tight', which are not defined terms.

The IK classification standard BS EN 50102 describes a system for classifying the degrees of protection provided by enclosures for electrical equipment against external mechanical impacts. The letters IK are followed by two numerals which identify a specific impact energy.

Appendix B of this Guidance Note lists the degrees of protection indicated by these classifications.

4.5 Presence of water (AD) 522

Any wiring system or equipment selected and installed must be suitable for its location and able to operate satisfactorily without deterioration during its working life. The presence of water can occur in several ways, e.g. rain, splashing, steam/humidity, condensation, and its effects must be considered at each location it is expected to be present. Suitable protection must be provided, both during construction and for the completed installation. 522-03

In damp situations and wherever they are exposed to the weather, all metal sheaths and armour of cables, metal conduit, ducts, trunking, clips and their fixings should be of corrosion-resistant material or finish. They should not be in contact with other metals with which they are liable to set up electrolytic action (see 4.7 also). 522-05

In any situation and during installation, the exposed insulation at terminations and joints of cables which are insulated with hygroscopic materials, e.g. mineral insulated cable, must be sealed against the ingress of moisture. Such sealing material, and any material used to insulate the conductors where they emerge from the cable insulation, should have adequate insulating and moisture-proofing properties and retain those properties throughout the range of service temperatures.

The manufacturer's recommendations regarding the termination and installation of all types of cables must be strictly observed.

In a damp situation, enclosures for cores of sheathed cables from which the sheath has been removed and for non-sheathed cables at terminations of conduit, duct, ducting or trunking systems should be damp-proof and corrosion-resistant. Every joint in a cable should be suitably protected against the effects of moisture. 526

Conduit systems not designed to be sealed should be provided with drainage outlets at any points in the installation where moisture might otherwise collect. Those outlets must not affect the electrical safety of the system, or allow vermin access. 522-03-02

If it is necessary to locate a switch, socket-outlet etc. near to a sink, the device must be protected from any water splashing that may occur. This can be achieved either by locating the device in a position where it will not be exposed to splashing, or providing a splash resistant covering, or providing a weather protected device. 512-06-01

It may occasionally be necessary to consider the installation of cables in locations that permanently contain water, such as fountains, marinas etc. It must be noted that the plastic insulating and sheathing materials are not totally waterproof and will absorb some water over time. Such cables are not suitable for continuous immersion and a cable manufacturer's advice should be taken for such installations. (See guidance Note 7 for further information.)

4.6 Presence of solid foreign bodies (AE) 522

Where dusty conditions may be expected to occur, equipment and enclosures for conductors and their joints and terminations should have the degree of protection of at least IP5X. There are two conditions recognised in this particular IP Code, the first condition being where the normal working cycle of the equipment concerned causes reductions in air pressure, e.g., thermal cycling effects, and the second 522-04 condition where such pressure reductions do not occur. These conditions should be considered during the design and selection procedure and appropriate equipment chosen.

It is necessary to have good operational housekeeping policies to ensure dust is kept to a minimum. A build-up of dust on electrical equipment, cables etc can act as thermal insulation and cause overheating of the equipment or cable. The dust may even be of a type that can ignite if the temperature of equipment rises. However, hazardous (explosive or flammable) dusts are outside the scope of this Guidance Note, and reference should be made to BS 6467 for further guidance.

To assist in cleaning in dusty installations, a design should be developed in which inaccessible surfaces and locations are minimised and surfaces that can collect dust are made as small as practicable, or sloped to shed dust. Cable ladder rack is preferred to cable tray, and ladder rack should be suspended vertically (i.e. on edge) rather than horizontally.

4.7 Presence of corrosive or polluting substances (AF) 522

In damp situations, where metal cable sheaths and armour of cables, metal conduit and conduit fittings, metal ducting and trunking systems, and associated metal fixings, are liable to chemical or electrolytic attack by materials of a structure with which they may come into contact, it is necessary to take suitable precautions against corrosion, such as galvanising, plating etc. 522-05

Materials likely to cause such attack include:

— materials containing magnesium chloride which are used in the construction of floors and walls

— plaster coats containing corrosive salts

— lime, cement and plaster, for example on unpainted walls, or over cables buried in chases

— oak and other acidic woods

— dissimilar metals liable to set up electrolytic action.

Application of suitable coatings before erection, or prevention of contact by separation with plastics, are recognised as effective precautions against corrosion.

Contact between bare aluminium conductors, aluminium sheaths or aluminium conduits and any parts made of brass or other metal having a high copper content, should be avoided especially in damp situations, unless the parts are suitably plated. If such contact is unavoidable, the joint should be completely protected against ingress of moisture. Modern, friction welded bi-metallic joints are available for most tape to tape or tape to cable joints. Wiped joints in aluminium sheathed cables should always be protected against moisture by a suitable paint, by an impervious tape, or by embedding in bitumen.

Bare copper sheathed cables should not be laid in contact with zinc plated (galvanised) materials such as cable tray etc. in damp conditions. This is because the electropotential series indicates that zinc is anodic to copper and therefore preferential corrosion of the zinc plating may occur. This action will not affect the copper, but may cause corrosion of the cable tray. The presence of moisture is essential to produce electrolytic action, therefore, in dry conditions this action will not occur. If moisture is present, then electrolytic action will take place, but the extent of any corrosion is dependent upon the relative areas of the two metals and the conductivity of the electrolyte (moisture in this instance). In these circumstances PVC sheathing should be used.

Hostile environments and chemicals also can attack the conductor, its insulation and sheath and any enclosure or equipment.

A few examples of this are:

— petroleum products, creosote, some solvents and hydrocarbons attack rubber and may attack polymeric materials such as pvc

— plasticisers migrate to polystyrene from pvc and also to some types of plaster (see 7.5 also)

— hostile atmospheres (e.g. those in the vicinity of plastics processing machinery, and vulcanising which produces sulphurous atmospheres)

— coastal areas with salt laden atmospheres

— in areas where animals are kept (agricultural environments, kennels, etc) animal urine can be corrosive.

In such cases the manufacturer's advice should be obtained and care taken in the measures employed.

To provide adequate protection from corrosion, the corrosive substances must be clearly identified and the manufacturer's or a specialist's advice obtained.

4.8 Impact (AG), vibration (AH) and other mechanical stresses (AJ) 522

Any part of the fixed installation which may be exposed to a severe impact must be able to survive it. In workshops where heavy objects are moved, the traffic routes should be avoided. If it is not possible to avoid the traffic routes heavy duty equipment or localised protection must be provided. 522-06 522-07 522-08

Similarly, a socket-outlet fixed at low level should not be placed where it is likely to be damaged by furniture or cleaning equipment. Conventionally socket-outlets are mounted at least 150 mm above the floor or a working surface.

See Appendix B for details of the IK code for classification of the degrees of protection by enclosures against impacts.

Final connections to plant that is adjustable or produces vibration must be designed to accommodate this and a final connection made in flexible cables or conduit or a properly supported cable vibration loop allowed.

Allowance must be made for the thermal expansion and contraction of long runs of steel or plastic conduit or trunking, and adequate cable slack provided to allow free movement. The expansion or contraction of plastic conduit or trunking is greater than that of steel for the same temperature change.

Cables crossing a building expansion joint should be installed with adequate slack to allow movement, and a gap left in any supporting tray or steelwork. A flexible joint should be provided in conduit or trunking systems. 522-12

4.9 Presence of fauna (AL), flora and/or mould growth (AK)

Special requirements for animal housing areas will depend on the type and size of animal and its required environment. Section 605 gives some guidance on farm requirements but specialist advice will be needed for more unusual species. 605

Cables and equipment may be subject to attack and damage from plants and animals as well as the environment. The damage may be caused by such diverse occurrences as vermin chewing cables, insect or vermin entry into equipment, physical impact/ damage by larger animals such as can occur in agricultural areas, and plant growth placing excessive strains on equipment over a period of time - a small lighting column base can be moved by tree roots - or choking equipment and blocking ventilation. Rodents have a particular taste for some forms of cable sheath and can gnaw through cables sheath and insulation to expose conductors. They build nests and the nests are usually constructed of flammable material. Such a combination is ideal for the propagation of fire where the nest surrounds the wiring material. Cables impregnated with anti-vermin compounds have not been found to be successful and may not comply with Health and Safety legislation. Cables should not be routed on likely vermin 'runs', e.g. on the tops of walls or in voids etc., but located in full view for easy inspection for damage. 522-09 522-10

As far as practicable, cables and equipment should be installed away from areas or routes used by animals or be of a type to withstand such attack. PVC sheathing and insulation may be chewed by vermin and in such areas steel conduit may be required.

The access of insects is difficult to prevent as they can enter through small gaps such as vent holes etc. Equipment and wiring systems in such locations must be carefully sealed and any vents fitted with breathers etc.

4.10 Potentially explosive atmospheres

The selection and erection of equipment for installations in areas with potentially explosive atmospheres of explosive gases and vapours or combustible dusts is outside the scope of this Guidance Note. The design, and the selection and erection of equipment for such installations requires specialist knowledge.

Further information may be found in BS EN 60079 - Electrical apparatus for explosive gas atmospheres, (parts 10 and 14 have replaced BS 5345), and BS 6467 'Electrical apparatus with protection by enclosure for use in the presence of combustible dusts.'

Section 5 — Installation of Cables

5.1 Cable selection 521

Cables must be selected to comply with the electrical characteristics of current rating, voltage drop etc. as required by Parts 4 and 5 of BS 7671 and the physical protection characteristics as required by Chapter 52. The correct selection and erection of a complete wiring system is necessary to provide compliance with Chapter 52.

Over-specification of cable performance requirements can result in increased installation costs, and the designer must assess performance reasonably. Cables must be selected to perform their required function in the environment throughout their expected life, but there is no justification for over design. (Consideration of the overall lifetime costs of a cable is not overdesign. There may be a trade off to increase the size of a cable to reduce overall energy losses in the cable during its life.) Regular periodic inspection (and testing as necessary) of an installation must be carried out to monitor cable and equipment conditions, and this should identify changes of use of an installation that would require modifications to the installation during its life. (See IEE Guidance Note 3 for further information).

Cables in emergency and life safety systems must be able to perform their intended function in emergency situations such as fire. BS 6387 "Performance requirements for cables required to maintain circuit integrity under fire conditions" specifies performance requirements and gives test methods for mechanical and fire tests applicable to cables rated at voltages not exceeding 450/750 V and for mineral insulated cables conforming to BS 6207. This standard specifies those requirements of the cables related to characteristics required to enable circuit integrity to be maintained under fire conditions.

The cables are intended to be used for wiring and interconnection where it is required to maintain circuit integrity under fire conditions for longer periods than can be achieved with cables of conventional construction.

Cables tested to this standard are categorised under three separate test conditions. A separate sample of cable can be used for each test.

The first letter indicates a resistance to fire alone (simulated burning):

A 650 °C for 3 hours
B 750 °C for 3 hours
C 950 °C for 3 hours
S 950 °C for 20 minutes.

The second letter **W** indicates resistance to fire at 650 °C with water, (simulating fire extinguishing for a 30 minute test period).

The third letter indicates resistance to fire with mechanical shock. Mechanical shocks are applied to the cable at specified temperatures of:

X 650 °C
Y 750 °C
Z 950 °C.

The highest cable rating is category CWZ which shows that a cable has passed the most severe test in each category, but not necessarily with the same cable sample. Designers must not assume that to specify any cable with a CWZ rating will provide total integrity in case of fire. Installation conditions and effects must be carefully considered. Cables may perform only for the stated time but may not be fire survivable, and could have only a limited integrity. Mineral insulated cable is the only cable understood to have passed the three tests of BS 6387 using the same sample of cable.

See Guidance Note 4 "Protection against fire" for further information on cable selection for fire protection and alarm systems.

BS 7540 : 1994 "Guide to use of cables with a rated voltage not exceeding 450/750 V" gives guidance on the general use of some types of cable, mainly the general wiring types (see Appendix F also).

General building wiring cables that comply with the requirements of BS 4066 Part 1 for flame propogation are considered to be of a suitable non flame propagating property to allow their installation in buildings and between fire zones or compartments without further physical fire protection unless there is significant risk of fire. (e.g. hazardous or flammable materials stored or special processes). Cables not to BS 4066 Part 1 may not pass unprotected between building fire compartments or zones and should generally only be used in short lengths and not in areas of fire risk.

522-12

527-01-04

Most types of general cable comply with BS 4066 Part 1, including PVC/PVC insulated and sheathed cables to BS 6004 and XLPE/SWA/PVC cables to BS 5467. A more severe test is prescribed in BS 4066 Part 3 which tests flame propagation of a group of cables and it may be that cables installed in groups in risers, or control and communications cables where in large volumes, should be required to comply with this standard.

TABLE 12 **Flame Retardant Cable Tests to IEC 332**

The two most widely recognised flame tests for flame retardant cables are IEC 332 Part 1 and IEC 332 Part 3 (identical to BS 4066 Part 1 and BS 4066 Part 3).

BS 4066 Part 1 A vertical flame test on a single wire or cable.	
BS 4066 Part 3 A flame test on bunched cables, 3.5 m in length, mounted on a vertical cable ladder. Part 3 has three distinct categories :	
	Category A (40 minute burn) - Large amount of cables on the ladder. (7 litres/metre of combustible material).
	Category B (40 minute burn) - Medium amount of cables on the ladder. (3.5 litres/metre of combustible material).
	Category C (20 minute burn) - Small amount of cables on the ladder. (1.5 litres/metre of combustible material).
The cables should not be affected/charred by the flame, beyond 2.5 m from the flame source.	

A new Standard, BS EN 50086 : Conduit systems for electrical installations - has been introduced for conduit systems which replaces:

BS 731 : Pt 1: 1952 (1993): Flexible conduits
BS 4568 : Steel conduit and fittings with metric threads
BS 4607 : Non-metallic conduit and fittings
BS 6099 : Conduits for electrical installations.

BS EN 50086 is in various parts; the following parts have been published:

BS EN 50086-1 : 1994	General requirements
BS EN 50086-2-1 : 1996	Particular requirements for rigid conduit systems.
BS EN 50086-2-2 : 1996	Particular requirements for pliable conduit systems.
BS EN 50086-2-3 : 1996	Particular requirements for flexible conduit systems.
BS EN 50086-2-4 : 1994	Particular requirements for conduit systems buried underground.

The BSs are still valid for conduit materials for repairs or modifications to older installations but the BS EN should be utilised for new installations.

BS EN 50085 : Part 1: 1998: Cable trunking systems and cable ducting systems for electrical installations - General requirements, has been published but the specific application standards are still being prepared. New BS ENs for cable trays and ladder systems are presently also being prepared.

These new standards are performance related and allow materials that will satisfy the required performance classification, and whilst these new Standards do not change any existing material specifications, they provide alignment of requirements throughout Europe. It is no longer correct to mention 'material specification' in relation to these standards. Products meeting the Harmonised Standards may then carry the 'CE' mark. However, it will be necessary to check the technical literature or with the manufacturer for special applications. 527-01-05

The designer must select an appropriate level of performance for application for the specific characteristics of the installation, and again, over specification in design will increase costs. A listing of performance criteria for conduit is given in Table 13 below.

TABLE 13
Performance requirements in BS EN 50086-1 : General conditions for conduit systems

Performance characteristics	Performance classification							
	0	1	2	3	4	5	6	7
Compression, N	0	125	320	750	1250	4000	-	-
Impact, N	-	0.5	1.0	2.0	6.0	20.0	-	-
Low temperature, °C	-	+5	-5	-15	-25	-45	-	-
High temperature, °C	-	+60	+90	+105	+120	+150	+250	+400
IP rating *Solid object*	-	-	-	2.5 mm	1.0 mm	dust protected	dust tight	-
Water	0	drops	drops at 15°	spray	splash-ing	jets	powerful jets	temporary immersion
Corrosion protection	-	low inside and outside	medium inside and outside	medium inside, high outside	high inside and outside	-	-	-
Tensile strength, N	0	100	250	500	1000	2500	-	-
Suspended load, N	0	20	30	150	450	850	-	-

Specific flame propagation tests are included in the conduit standards, although there is no conduit, trunking or tray that has a 'low emission of smoke and corrosive gases when affected by fire'. Flame propagation is defined in terms of a pass or fail test based on temperatures consistent with human safety (temperatures above which persons cannot survive). Materials that pass the test are identified as "non-flame propagating" and are required to be marked with a clearly legible identification or may be of any colour except yellow, orange or red. Flame propagating conduit must be coloured orange. Related fire products such as smoke production are being considered, but as yet are not covered in the Standards. 527-01-05

5.2 Cable concealed in structures 522

Where a cable is to be concealed in plaster, capping or conduit may be provided to position the cable and to reduce the risk of damage during the plastering process. 522-06

BS 7671 requires that non-sheathed cables for fixed wiring shall be in conduit, ducting or trunking. Non-sheathed cables must not be installed in building ducts formed in-situ because drawing-in of such cables is likely to damage their insulation. 521-07-03

Cables to be installed on walls and the like should have a sheath and/or armour suitably resistant to any mechanical damage likely to occur or be contained in a conduit system or other enclosure able to guard against such damage. 522-08

Cables concealed within a wall must be carefully routed. The route and depth must be such that the cables are not in an unexpected place. Fixings for other items sharing the space must be positioned so that there is no risk of penetrating the cables. BS 7671 details conventional routes within which there is a high risk of encountering a cable (see Fig 5). 522-06

Where a cable is concealed in a wall or partition at a depth of less than 50 mm it must be installed either horizontally within 150 mm of the top of the wall or partition or vertically within 150 mm of the angle formed by two walls or partitions, or run horizontally or vertically to an accessory or enclosed in earthed metal conduit, trunking or ducting or have an earthed metallic sheath if it is run outside these areas (see Fig 5). It must be remembered that for walls or partitions of which both sides are accessible e.g. internal walls within a building, cables installed vertically or horizontally from an outlet on one side of the wall must be at least 50 mm clear of the other side or protection will be required. This may be a problem in newer housing construction, where walls or partitions may be thinner than buildings of traditional construction. 522-06-06 522-06-07

Fig 5: **Permitted cable routes**

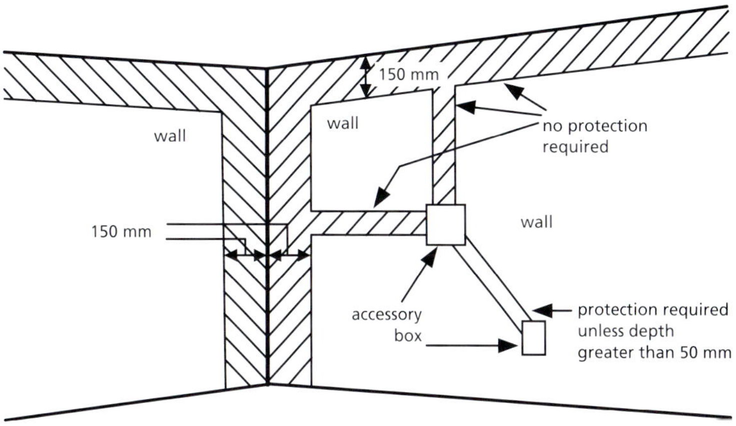

Where a cable is concealed within a wall or partition and cannot be run in the prescribed areas, it is to incorporate an earthed metal sheath or be protected by an earthed metal covering. Any penetration of a phase conductor through the metal will then create a fault current to earth and operate the protective device. 522-06-06 522-06-07

If a cable is installed under a floor or above a ceiling it shall be run in such a position that it is not liable to damage by contact with the floor or ceiling or their fixings. Cables passing through a joist shall be at least 50 mm from the top or bottom as appropriate or enclosed in earthed steel conduit or have an earthed metallic sheath (see Fig 6). 522-06-05

Regulations 522-06-05 and 07 allow, as an alternative to earthed metal conduit, '*mechanical protection sufficient to prevent penetration of the cable by nails, screws and the like*'. This protection is a very difficult bearing in mind the modern fixing such as self-tapping screws and shot fired nails that are available. In wood floors where the nail or screw could be driven into the floorboard as a guide before being driven through a cable into a joist, the use of metal covers at joists may not be adequate. Also, Regulation 522-06-05 requires cables to be at least 50 mm from the top or bottom of a joist. It is good practice to ensure that the 50 mm is maintained along the whole route as nails or screws are a similar size all along the route. 522-06-05 522-06-07 522-06-05

Regulation 522-12-03 requires that penetrations for wiring must not be made in any load bearing (or 522-12-03 structural) element of a building unless the integrity of the element can be assured after the penetration. Such assurance is outside the ability of most contractors so great care must be taken to plan cable routes and specialist structural advice is always desirable.

It is accepted that there may be a structural problem in older buildings if sections of floor joists continue to be removed for rewiring, plumbing, etc and it may be necessary to reuse existing joist penetrations. The designer and installer will then have to develop a system that complies with the requirements of the IEE Wiring Regulations (BS 7671).

The National House Building Council (NHBC) publish advice in their standard on internal services on the notching and drilling of timber joists and this information is reproduced below for information. It is emphasised that any notching, cutting or drilling should be designed with the approval of a competent structural engineer. (522-12-03) 522-12-03

Timber joists and studs should only be notched and drilled within the limits shown in the table below:

TABLE 14 **Joist and stud notch and drill limits**

Item	Location	Maximum size
Notching joists up to 250 mm depth	top edge 0.1 to 0.2 of span	0.15 x depth of joist
Drilling joists up to 250 mm depth	centre line 0.25 to 0.4 of span	0.25 x depth of joist
Drilling studs	centre line 0.25 to 0.4 of height	0.25 x depth of stud

Fig 6: **Holes in joists**

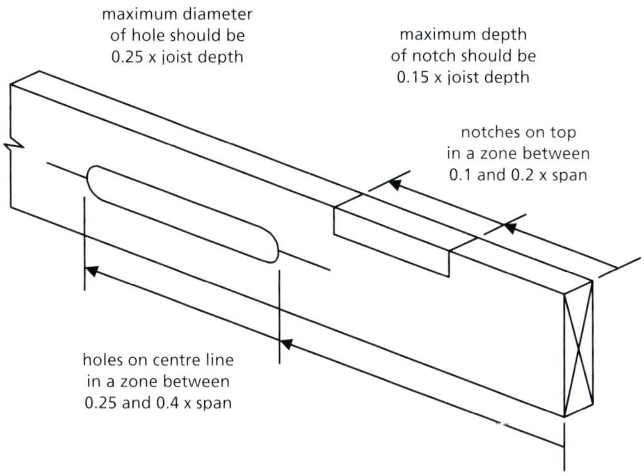

Notches and drillings in the same joist should be at least 100 mm apart horizontally.

It should be noted that the NHBC guidance has slight differences from The Building Regulations approved document A1, which recommends that notches should be no deeper than 0.125 times the joist depth and in a position between 0.07 to 0.25 of the span and holes should be of no greater than 0.25 times the joist depth and drilled in the neutral axis of the joist.

Fig 7: **Cables in floor joists**

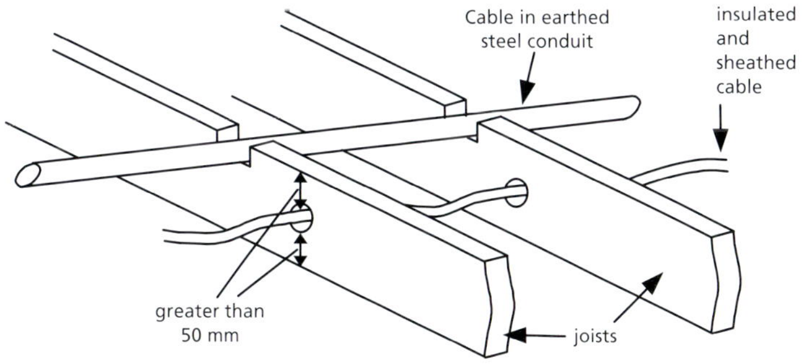

5.3 Cable routes and livestock 605

Particular attention is needed in locations where livestock is present. (See 4.9 also). 605

Animals in enclosures or buildings may rear up or move as a herd. Such movement can result in severe damage to inappropriately positioned equipment and risks to the animals themselves. Animals may also create corrosive conditions so that normal finishes are not durable enough. Heavy-duty equipment with anti-corrosive finishes should be selected and appropriately sited. Class II installations in heavy duty high impact pvc with suitable Class II accessories and conduit and enclosures have been found to be satisfactory in many cases.

5.4 Capacity of conduit and trunking 522

Where cables are installed in conduit or trunking care is needed to prevent undue strain when installing the cables. A method for dealing with this is given in Appendix A. This may be used to determine the minimum size of conduit or trunking necessary to accommodate cables of mixed sizes. 522-08

The method employs a 'Factor System', each cable size being allocated a convertible factor. The sum of all factors for the cables intended to be run in the same enclosure is compared against the factors given for conduit or for trunking, to determine the minimum size of conduit or trunking necessary to accommodate those cables. The derivation of the tables is based on practical tests involving an easy draw of cables into the conduit or trunking.

It must be remembered that cables installed in conduit or trunking must be sized to allow also for the thermal effect of the grouping of such cables. Where advantage is taken of grouping factors it may be necessary to indicate this by a local label in the trunking or a note in the installation record documentation to warn future modifiers.

5.5 Selection of cables and wiring systems with low emission of smoke and corrosive gases when affected by fire 52

When specifying or selecting cables in areas with high levels of public access such as hospitals, schools, supermarkets, shopping developments etc., the designer should consider factors which may affect the safety of the occupants, such as:

(i) where the rapid formation of dense noxious smoke would be likely to produce injury or panic in individuals or crowds

(ii) along escape routes, where it is important that the rapid but orderly exit of people is not impeded by smoke or fumes

(iii) where the formation of corrosive halogen-acid products would have a harmful effect upon expensive equipment, i.e. computer hardware.

A list of British Standards is given in Appendix 1 of BS 7671 and includes reference to cables with low emission of smoke and corrosive gases when affected by fire. These include BS 6724, BS 7211 and BS 7629.

Unplasticised pvc (pvc-u) as used in the manufacture of many cable management products (conduit, ducting, trunking and cable tray) is non-flame propagating and is inherently fire-resistant due to its halogen content.

It is not expected that general electrical cables in the quantities installed in most buildings would provide high levels of smoke on their own in a fire, and cables would usually only be expected to burn when other building components are also on fire. However, in some installations such as public buildings, hospitals etc., it is preferable to take positive action to limit the production of smoke and fumes. Cables with low emission of smoke and corrosive gases when affected by fire have an important place in the installation design.

In general, cables with low emission of smoke and fumes are split into either thermoplastic or elastomeric. Elastomeric cables with low emission of smoke and fumes include EPR, EVA and SILICONE and are extruded on rubber CV (continuous vulcanisation) lines. Thermoplastic cables with low emission of smoke and fumes are processed like pvcs and each cable company will have different compound formulations and processing. Generally, cables with low emission of smoke and fumes are flame retardant (with a high oxygen and temperature index), exhibit low smoke and fume emission with low (or zero — expressed as < 0.5 %) HCL (hydrochloric) acid gas emission. PVC compounds do not come under the cables with low emission of smoke and fumes category.

Oxygen index (given as limiting oxygen index or LOI) defines the minimum percentage of oxygen required in an oxygen/nitrogen test atmosphere to just support combustion. A higher figure identifies a more flame retardant cable. This index is not however fully representative of real fire conditions and the more recently established temperature index test gives a better prediction of a material's behaviour.

The temperature index is defined as the temperature at which combustion of the material under test is just possible at a normal atmospheric oxygen concentration. Again, a higher index indicates a better performance.

Tests for smoke and acid gas emission identify the performance of cables in fire conditions. These tests need to take into account the combination effect of all the materials used in a wiring system and environmental conditions. Specific requirements for smoke and acid gas limits must be specified by the designer and statutory authorities where this is relevant. It must be realised that needless over specification of cables with low emission of smoke and fumes will impose an unnecessary cost burden on a project.

Buildings required to comply with the special provisions of "Section 20" in London, and equivalent provisions in other cities, may require special precautions. Cables with a low emission of smoke and corrosive gases when affected by fire or steel conduit and trunking installations may be necessary when cables are installed in building voids, such as suspended ceiling voids, where these are used as plenums. Such design requirements should be clarified with the local Fire Officer or Building Control Officer before design commences.

Section 20 of the London Building Acts (Amendment) Act 1939 (as amended primarily by the Building (Inner London) Regulations 1985) is principally concerned with the danger arising from fire within certain classes of buildings which by reason of height, cubic capacity and/or use necessitate special consideration. The types of building coming within these categories are defined under Section 20 of the amended 1939 Act.

As buildings vary so much in height, cubic capacity, layout, siting, use and construction the relevant Council will deal with each case on its merits. The principles seek, in certain buildings (or parts of buildings) to ensure the safety of the structure against fire.

Section 20 applies where:

(a) a building is to be erected with a storey or part of a storey at a greater height than:

(i) 30 metres, or

(ii) 25 metres if the area of the building exceeds 930 square metres

(b) a building of the warehouse class, or a building or part of a building used for the purposes of trade or manufacture, exceeds 7,100 cubic metres in extent unless it is divided by division walls in such a manner that no division of the building is of a cubical extent exceeding 7,100 cubic metres.

A fire alarm system complying with the current edition of BS 5839 Pt 1 should be provided throughout every building except buildings comprising flats and/or maisonettes. These are covered by BS 5839 Pt 6 and also may have specific Building Regulations requirements.

In some instances, it may be necessary for such a fire alarm installation to be automatic where the use of the building (or part) warrants it (e.g. in hotels).

In office buildings where the means of escape is based on phased evacuation a number of additional features (such as a public address system) will be necessary.

The fire brigade when responding to an emergency in a building should have the indicator panels and associated manual controls for the building's fire protection systems located together in one place designated as the fire control centre.

Smoke extraction should be provided from each storey by openable windows or by a mechanical smoke extract system.

Diesel engine generators or pumps, should be enclosed by walls and a roof, of non-combustible construction having a standard of fire resistance of not less than four hours. Any openings in the walls should be protected by a single self-closing 120 minute door.

Oil storage in connection with the foregoing should comply with the fire brigade requirements.

Transformer substations and switchrooms containing electrical oil-cooled transformers or oil-filled switchgear with an oil capacity in excess of 250 litres, will be required to be housed in a fire resistant substation area. Cast resin, or dry-type transformers may be a viable option, although initially more expensive.

Conditions will normally be imposed requiring details of the general heating, lighting, electrical and ventilating arrangements to be submitted to the local Council for approval before commencement of the works and for such works to be provided and maintained to the satisfaction of the local Council, and periodic inspections may be made by local Council inspectors of the approved heating, lighting, electrical and ventilation installations during installation.

The foregoing Section 20 requirements apply specifically to London, but other large cities have their own similar requirements.

5.6 Buried cables

Cables to be installed in ducts or pipes in the ground are not required by BS 7671 to have armour, if the duct or pipe is sufficiently strong to resist likely mechanical damage, but in case of doubt, armouring should be provided. 522-06-03

Cables other than earthing conductors buried directly in the ground are required to have earthed armour or metal sheathing or both, or be of the concentric type with a tough sheath. 522-06-03 542-03 611-04

All buried cables must be marked by cable covers or a suitable warning tape, and buried conduits or ducts must also be suitably identified. A warning tape above the duct or conduit run would be suitable. 522-06-03

Cables, conduits and ducts must be buried deep enough to avoid being damaged by any disturbance of the ground reasonably likely to occur. This could include gardening/horticultural works, excavating for any local buried service pipes etc. Generally it would be better to locate cables etc. clear of such possible works.

A depth of burial of less than 500 mm is usually inadvisable as shallow laid cables may be inadvertently damaged by general gardening etc. Cables that cannot be buried at a reasonable depth should be specifically protected, e.g. by ducts encased in concrete, or installed along an alternative route.

Before any excavation is undertaken for cable or other works the HSE guidance booklet HS(G)47 "Avoiding danger from underground services" should be studied as it provides valuable advice on safety aspects.

It is important to be able to identify exactly where hidden services are located and accurate records, including drawings, should be made before trenches are backfilled. Cable routes should be located on drawings by dimensions from some fixed object that is not expected to change for a considerable time, (e.g. a building). Road and pathway features may be relatively easily changed. Scale drawings are of little value unless very detailed. Concrete route markers may be installed at changes of direction of cable routes and regularly along routes. It may also be useful to lay cable warning marker tape just below ground level along cable routes so this will be exposed at any future digging well before cables are accessed. 514-09-01

Many public service undertakings with services buried along the public highway have agreed in The National Joint Utilities Group (NJUG) a specific colour identification system for their service or duct work. Generally each of the Statutory Undertakings has been allocated a specific colour for their underground services. Details are given in Table 15.

In addition, Local Authorities and in particular Highway Authorities have allocated themselves colours appropriate for their own installations. These vary from area to area and on occasion may clash with the NJUG colour code. In the interests of safety, Local Authorities should be asked to supply details before excavation is commenced in their area.

TABLE 15

Agreed colours of ducts, pipes cables and marker/ warning tapes for public service undertaking

(Prepared by National Joint Utilities Group (NJUG) detailed in their publication No 4)

	Colour of duct/pipe/cable buried in ground			**Colour of marker/warning tape**
Utility	**Duct**	**Pipe**	**Cable**	**where used**
Gas	Yellow (or pale green)	Yellow	—	Yellow with black legend
Water	Blue	Blue	—	Blue
Electricity	Black	—	Black (Red for some HV)	Yellow with blue legend
BT (British Telecommunications) plc	Light grey	—	Light grey and black	White with blue legend
Mercury Communications Ltd (MCL)	White	—	Black All cable installed in duct	White with blue legend
Cable TV and telecommunications services other than BT and MCL	Green	—	Black All cable installed in duct	Green and/or yellow with identification showing co-axial or optical fibre cable
Highway Authority Services	**Duct**	**Pipe**	**Cable**	**Tape**
Street lighting England and Wales	Orange	—	Black	Yellow with black legend
Street Lighting Scotland	Purple	—	Purple	Yellow with black legend
Traffic control	Orange	—	Orange	Yellow with black legend
Telecommunications	Light grey	—	Light grey (or black)	Yellow with black legend
Motorways England and Wales				
Communications	Purple	—	Grey	Yellow with black legend
Communications power	Purple	—	Black	
Road lighting	Orange	—	Black	
Scotland				
Communications	Black or grey	—	Black	Yellow with black legend
Road lighting	Purple	—	Purple	

Notes:

1. The NJUG code is not retrospective and older installations installed before the code was adopted may not conform to the above colour scheme.
2. Cables installed on private property may not follow the guidelines in this document. The owner concerned should be contacted for information on the colour code applying in these cases.
3. Abandoned ducts are sometimes used for other purposes. Ensure systems are fully identified and traced out before any work is carried out.

The colour code is not universally applied as many of the older services were installed before this allocation was developed. Other means of identifying cables include clay or concrete tiles laid above electricity cables, usually with the word 'electricity' embossed into the top surface. The use of coloured plastic tape or board which would have the details of the service printed on its surface is a more recent method.

Further guidance may be obtained from the National Joint Utilities Group or other specifiers (see Appendix L), or any public services supplier.

Where cables pass through holes in metalwork, brickwork, timber, etc., precautions must be taken to prevent damage to the cables from any sharp edges, or alternatively mechanical protection should be provided, sufficient to prevent abrasion or penetration of the cable. 522-08

5.7 Sealing and fire stopping

Where a cable, conduit or trunking passes through a building wall or floor slab it is necessary to seal the hole around the cable etc. to the standard of the original wall or slab to prevent the passage of moisture, condensation, vermin or insects etc. The sealant must be suitable for the cable etc. and the environment and should be flexible to allow some small movement of the cable etc. relative to the building. Seals should be installed when necessary to maintain the building integrity during construction. 527-02-01 527-02-03 527-03-01

Where the penetration is through any part of the construction identified as having a specified fire performance or fire resistance the sealing must maintain the required fire integrity. The sealing method must be flexible and should be specified by the installation designer in liaison with other design team members such as the architect. The sealing must be properly installed through the thickness of the wall or slab, and inspection of the installation of the sealing is required by Regulations 527-04-01, and 712-01-03 item (vii). It is unlikely however that the electrical installer will be able to determine the suitability of a particular sealing method and specialist advice should be sought. 527-02-02 527-04-01 712-01-03

Conduits, trunking and ducts passing through the building elements have space inside them that can conduct combustion products or other harmful materials (e.g. explosive gases and vapours etc.) and may need to be sealed internally also. Regulation 527-02-02 states that a non flame propagating wiring system (see 5.1) with a maximum internal cross sectional area 710 mm (i.e. 25 mm diameter conduit or smaller or 25 x 25 trunking) does not require to be internally sealed for fire protection. It may however still require to be sealed for other reasons such as condensation etc.

527-02-02

Section 6 — Sizing of Cables

6.1 Current-carrying capacity and voltage drop 525

The Preface to the current rating and voltage drop tables of Appendix 4 of BS 7671 advise how they should be used to work out the cross-sectional area of a conductor. This will depend, among other factors, on the type of overcurrent protection provided. Guidance Note 6 gives further information on cable sizing. Appx 4

Although the nominal supply voltage in the UK is now 230/400 V (+6%-4% tolerance), in practice, there will be no change in the voltage supplied. Therefore, this guidance document has used 240/415 V for some calculations.

The design current of a circuit should be that drawn by the equipment at the nominal voltage of the supply.

6.2 Diversity 311

Diversity at any point in time is the actual load (or estimated maximum demand) divided by the potential load (or connected load), usually expressed as a percentage. The load that may be taken for the basis of selection of cable and switchgear capacities, allowing the design diversity, must be the maximum foreseeable load figure that will occur during the life of the installation. 311-01-01

The maximum connected load is usually the simple sum of all the electrical loads that are, or may be connected to the installation. The electrical load of each item will be the maximum electrical load that item may require, such as the load of an electric motor, unless specific measures have been taken to ensure that, in a group of items with high inertia starting loads, or inrush currents, such loads cannot occur simultaneously. Frequent motor starting will require special consideration.

A safe design diversity would be 100 per cent of the potential (or connected) load but this clearly is not reasonable. This is recognised, for example, in Appendix C of this Guide, where an unlimited number of 13 A socket-outlets may be connected to a single 30 A or 32 A ring final circuit. If there are twenty 13 A socket-outlets (each capable of supplying approximately 3 kW at 230 V) the potential load of the circuit is 60 kW but the maximum permissible for a 30 A circuit is 6.9 kW, a diversity of 11.5 per cent. In practice the actual load on a dwelling ring final circuit is usually well below 6.9 kW except perhaps in a kitchen or utility room.

The economic design of an electrical installation will almost always mean that diversity has to be allowed. However, where there is doubt as to the factors to be used, an adequate margin for safety should be allowed and consideration should also be given to the possible future growth of the maximum demand of the installation.

Section 433 requires circuits to be so designed that small overloads of long duration are unlikely to occur. 433-01-01

For a guide towards the estimation of diversity see Appendix H.

6.3 Cross-sectional areas of conductors 524

Table 52C of BS 7671 considers only the minimum cross-sectional areas of conductors that should be used. The actual cross-sectional area chosen depends on a number of factors. These are dealt with in considerable detail in the Preface to the tables of Appendix 4 of BS 7671. 524-01 Appx 4

Some loads such as motors, transformers and some electronic circuits take large surge (in-rush) currents when they are switched on. Due account must be taken of any extra cumulative heating effect due to motor in-rush currents or electric braking current where the motors are for intermittent duty with frequent stopping and starting. 552-01-01

Consideration must be given to the cross-sectional areas of neutral conductors. In a single phase circuit the neutral conductor shall be at least the same cross-sectional area as the phase conductor. With circuits 524-02-01 for discharge lighting a reduced neutral conductor is not permitted, but generally Regulation 524-02-02 allows the possibility of a reduced size neutral in polyphase systems where the neutral conductor will not be overloaded (523-01-01). Even with balanced three-phase conditions there can be high percentages of third harmonic currents which add arithmetically in the neutral conductor, and when assessing the maximum likely neutral current allowance must be made for any unbalanced loading or harmonics.

524-02-02

523-01-01

Where the circuit is for discharge lighting the design current, in the absence of more exact information, can be taken as:

$$Ib = \frac{1.8 \text{ x } \textit{lamp rated wattage}}{\textit{nominal voltage of circuit}} \text{ amps}$$

The multiplier (1.8) is based on the assumption that the circuit is corrected to a power factor of not less than 0.85 lagging. It takes into account controlgear losses and harmonic currents. If corrected discharge lighting is used the power factor will be typically 0.85 lag but uncorrected discharge lamps may vary between 0.5 lead and 0.3 lag.

High neutral currents may be encountered with information technology equipment using switch-mode power supplies. In such equipment the mains input is rectified, the d.c. output being fed to a capacitor which in turn supplies the switch-mode regulator.

It has been suggested that the rms current in the neutral conductor of a three-phase mains circuit can be 1.73 times the rms current in the phase conductors.

Thus when designing three-phase circuits for equipment incorporating switch-mode power supplies the designer may need to determine from the equipment manufacturer the rms current taken by the equipment and the expected neutral current and its harmonic content. It may sometimes be necessary to install a neutral conductor of larger size than the associated phase conductor.

BS 7450 : 1991 "Method for determination of economic operation of power cable size" based on overall lifetime costs (including cable and installation costs and the cost of waste energy) may also be of assistance in the determination of optimum conductor sizes based on energy losses.

6.4 Voltage drop in consumers' installations 525

Consideration must be given to circuit voltage drop in both the steady state and transient situations. The requirements of the steady state condition are dealt with in BS 7671. Under transient conditions e.g., during motor starting periods, a greater voltage drop may be accepted, but BS 7671 requires that the voltage variations should not exceed those specified in the appropriate British Standard or, in the absence of a British Standard, the manufacturer's recommendation. 525-01 525-01-01

A number of British Standards prescribe tests intended to prove that the equipment concerned operates safely at other than its rated voltage.

When designing an installation it should be borne in mind that the suppliers are allowed, under the Electricity Supply Regulations 1988 (as amended) a +10 per cent to -6 per cent variation on the declared nominal voltage of 230 V.

The designer may take the approach stated in Regulation 525-01-02, but must bear in mind that a 5 per cent reduction in voltage may result in a 10 per cent reduction in power output, and an even larger drop in the light output of incandescent lamps. Data on the performance and life expectancy of incandescent lamps can be obtained from manufacturers. 525-01-02

Section 7 — Other Influences

7.1 Electrical connections 526

Connections between cables and other equipment must be selected with care. There is a risk of corrosion between dissimilar metals depending upon the environment. The manufacturers' recommendations must be observed and the Standard followed. The power rating and the physical capacity of terminals must be adequate, as must their physical strength, when terminating larger conductors. 526 522-05

XLPE insulated cables to BS 5467 can operate at higher conductor temperatures than pvc-insulated cables. The higher conductor temperature results in an increased current-carrying capacity (10 to 20 per cent) and higher voltage drop values.

The use of XLPE insulated cable requires careful selection of termination systems and accessories and switchgear which are appropriate to the operating temperature of the cable. Alternatively, where XLPE insulated cable is to be used, it should be derated (its operating current I_z reduced) so as not to allow the conductor temperature to exceed the maximum operating temperature of the equipment to which it is to be connected. Conversely, XLPE insulated cables may be a necessity where the heat of the device (e.g. busbars) may be excessive. PVC or XLPE conductors used in this way may need their insulation locally replaced or supplemented by an insulation with a more suitable temperature rating. Where it is necessary to adjust the cross-sectional area of a cable at the point of termination this should be done by the use of a recognised technique such as a butt splice or crimp connector. 512-02-01 523-01-01

Any enclosure must have suitable mechanical and fire resistance properties. Unless exempt, joints must remain accessible for inspection and maintenance. 526-03 526-04

Account must be taken of any mechanical damage and vibration likely to occur. The method of connection adopted should not impose any significant mechanical strain on the connection, nor should there be any mechanical damage to the cable conductors. 526-02 526-03-01

Mechanical, clamps and compression-type joints should retain securely all the strands of the conductor. Every compression joint should be of a type which has been the subject of a test certificate as described in BS 4579, and the appropriate tools and methods specified by the manufacturers of the joint connectors should be used. A termination or joint in an insulated conductor, other than a 4 sq mm or greater protective conductor, should be made in an accessory or luminaire complying with the appropriate British Standard. Where this is not practicable it should be enclosed in material designated non-combustible to BS 476, Part 4 or complying with 'P' to BS 476, Part 5 or the relevant requirements of BS 6458, Section 2.1. Such an enclosure may be formed by part of an accessory and/or luminaire and a part of the building structure. 526-01-01

All connections, including those of ELV 12 V luminaires, must be enclosed in accordance with the requirements of Regulations 422-01-04 and 526-03. The use of unenclosed or taped up terminal blocks or strips, which themselves may not be tested to any recognised standard, does not comply with the requirements of this Regulation unless the connections are enclosed within a properly constructed junction box or enclosure either to the relevant British Standard (e.g. BS 4662 or BS 5733) or made with a material which passes the glow wire test of BS 6458, Section 2.1, or is within an enclosure formed from a suitable building material or formed within the building structure. Such building material must be non-combustible, or structure part must have suitable ignitability characteristics when tested as specified in BS 476. All insulated, unsheathed cable, such as cable ends stripped back, must be contained in the enclosed. 422-01-04 526-03

Terminations of mineral-insulated cables should be provided with sleeves (tails) having a temperature rating similar to that of the seals.

Cable glands should securely retain without damage the outer sheath or armour of the cables. Mechanical cable glands for rubber and plastics insulated cables should comply with BS 6121 where appropriate.

Appropriate cable couplers should be used for connecting together lengths of flexible cable or flexible cord. Generally cable couplers must be non-reversible and to an appropriate Standard. On construction sites, where the requirements apply, every plug and socket-outlet must comply with BS 4343 (BS EN 60309-2).

553-02-01
604-01-01
604-12-02

Conduit should be free from burrs and swarf and all ends should be reamed to obviate damage to cables.

Substantial boxes of ample capacity should be provided at every junction involving a cable connection in a conduit system. The designer should be aware that large cables need a greater factor of space for installation and removal.

Regulation 460-01-06 requires a provision for disconnecting the neutral conductor for isolation or testing purposes. A suitable joint will be adequate for smaller cables, but a bolted link could be used for larger conductors due to the physical effort necessary to move the cables for disconnection.

460-01-06
537-02-05

7.2 Cables in contact with thermal insulation 523

Where a cable is to be run in a space in which thermal insulation is likely to be placed, the cable shall wherever practicable be fixed in a position such that it will not be in contact with or covered by the thermal insulation. Where fixing in such a position is impracticable the cross-sectional area of the cable shall be appropriately increased.

For a cable installed in a thermally insulated wall or above a thermally insulated ceiling, the cable being in contact with a thermally conductive surface on one side, current-carrying capacities are tabulated in Appendix 4 of BS 7671, Method 4 being the appropriate Reference Method.

For single cable likely to be totally surrounded by thermally insulated material over a length of more than 0.5 m, the current-carrying capacity shall be taken, in the absence of more precise information, as

0.5 times the current-carrying capacity for that cable clipped direct to a surface and open(Reference Method 1).

Where a cable is totally surrounded by thermal insulation for less than 0.5 m the current-carrying capacity of the cable shall be reduced appropriately depending on the size of cable, length in insulation and thermal properties of the insulation. The derating factors in the table below are appropriate to conductor sizes up to 10 mm^2 in thermal insulation having a thermal conductivity greater than 0.0625 W/Km.

TABLE 16
Cables surrounded by thermal insulation

Length in insulation (mm)	Derating factor
50	0.89
100	0.81
200	0.68
400	0.55
500 and over	0.50

The current-carrying capacities are the same for cables installed in the following manner (Reference Method 1): Table 4A

(i) sheathed cables clipped direct or lying on a non-metallic surface (Installation Method 1).

(ii) sheathed cables embedded directly in masonry, plaster and the like (Installation Method 2)

For cables in conduit in a thermally insulating wall or above a thermally insulating ceiling where the conduit is in contact with a thermally conductive surface on one side (Method 4) the current-carrying capacities are lower than for Method 1. 523-04

No current-carrying capacities for cables totally surrounded by thermal insulation are tabulated in Appendix 4 of BS 7671. In the absence of more precise information the current-carrying capacities of cables totally surrounded by thermal insulation over a length of more than 0.5 m should be taken as half that for cables clipped direct and derating factors are given for lengths shorter than 0.5 m. Table 52A is based Table 52A 523-04 upon normal cavity insulation such as vermiculite granules. More exact data can be obtained from ERA Report 85-0111 'The temperature rise of cables passing through short lengths of thermal insulation'.

The following notes consider the effect of such derating on various typical circuits.

For domestic lighting circuits, the load does not normally exceed 6 A. The smallest cable used will be 1 mm^2 with a current-carrying capacity for Method 1 of 15 A (Table 4D2A, Column 6). If totally surrounded by thermal insulation for 0.5 m or a greater length the current-carrying capacity can therefore be taken as 7.5 A. Appx 4

The cable most frequently used for cooker circuits is 6 mm^2 which has a current-carrying capacity, when clipped direct, of 46 A, reducing to 23 A if it is installed in thermal insulation. A 30 A or 32 A overcurrent device is usually employed. The current taken by a cooker varies over a wide range and theoretical loads in excess of 23 A can be expected. This would lead to overheating of the cable if continuously carrying such current. Hence a larger cable may be necessary. Also, the nominal current (I_n) of a protective device must not exceed the lowest of the current carrying-capacities of any of the circuit conductors (I_z). 433-02-01(ii)

The loading of cables feeding socket-outlets is even more varied, and depends on the type of circuit used, i.e. ring or radial, the number of socket-outlets fed from the circuit and the type of load. For example, 2.5 mm^2 pvc insulated and sheathed cables used in a multi-socket outlet radial circuit protected by a 20 A overcurrent protective device has a 'clipped direct' current-carrying capacity of 27 A (Table 4D2A, Column 6) reducing to 13.5 A if the cable is embedded in thermal insulation.

When clipped direct the cable could supply up to 6.5 kW without overheating, though because, in this example, the circuit is protected by a 20 A device it should not be loaded to more than 4.8 kW in any event.

Loading as great as this is infrequently encountered and much more common are small loads such as standard lamps and television sets, up to about 1 kW, i.e. 4 A in total.

However, if the cable is later embedded in thermal insulation the reduced current-carrying capacity of 13.5 A would be exceeded if a load of 3.2 kW were connected and overheating is possible.

Thus, the introduction of thermal insulation into cavities where cables are already installed will create the possibility of overheating the cable. The extent of the risk will depend on the type of thermal insulation used and the total loading of the cables.

7.3 Mutual or individual deterioration 522

BS 7671 requires that materials liable to cause mutual or individual deterioration or hazardous degradation shall not be placed in contact with each other. One commonly used thermal insulation for cavity walls is urea formaldehyde foam and there should be no adverse action between this material and pvc. 522-05-03

Other materials may be used for cavity insulation but the supplier does not always disclose their chemical composition. Before allowing any thermal insulation materials to come into contact with cable insulation or sheath materials the thermal insulation supplier should be asked to confirm in writing that there will be no adverse effect on the cable insulation or sheath. Where there is doubt an inert barrier should be inserted between the cable and the thermal insulation.

Expanded polystyrene sheets or granules are used in construction, e.g. insulating lofts and supporting floors, and may be used in cavity walls. If this material comes into contact with pvc cable sheathing some plasticiser will migrate from the pvc to the polystyrene. The pvc will become less flexible and sticky on the surface, and the polystyrene will become soft and shrink away from the pvc sheathed cable if possible. Such contact between pvc and polystyrene should be avoided. These comments do not apply to unplasticised pvc conduit and trunking systems.

As only the plasticiser is removed from the sheath, it is not expected that the insulation resistance of the cable will be affected if the cable is not disturbed, but this cannot be guaranteed. The cable should be supported away from the polystyrene and a test should be carried out. The cable condition should also be monitored over time in regular periodic inspections. There is no cure for this migration other than replacement of the cable.

Similar remarks apply to some fittings which are in contact with cables. In addition to polystyrene and expanded polystyrene, acrylonitrile-butadiene-styrene (ABS), polystyrene and polycarbonate are also affected. Nylon, polyester, polyethylene, polypropylene, rigid pvc and most thermosetting plastics are little affected. Natural rubber grommets can become softened but synthetic rubber and pvc grommets are satisfactory.

Contrary to popular belief, plastic insulated and sheathed cables are not suitable for continuous immersion in water, such as in pools and fountains, as they will absorb an amount of water over time and insulation failure will occur. A manufacturers' advice should be taken for applications involving continued immersion of cables. Cable standards for these purposes are currently being developed.

7.4 Proximity to other services 528

Electrical and all other services must be protected from any harmful mutual effects foreseen as likely under conditions of normal service. 515-01 528-02

In the second amendment to BS 7671: 1992 (AMD 9781) the references to the three categories of circuit used in Section 528 has been deleted and replaced with a reference to two voltage bands, equating approximately to ELV and LV. This change was introduced to simplify system separation as BS 7671 previously identified several separation requirements. Care must be taken that only compatible circuits are enclosed in the same conduit or trunking, unless the circuit conductors are insulated to the highest voltage present, or other precautions are taken, such as segregation by a separate compartment in an insulated cable system or by an earthed metallic screen. 528-01-03

For safety circuits, such as emergency lighting and fire alarms, covered by BS 5266 and BS 5839, the 528-01-04 requirements of the British Standards must be met with respect to segregation. For telecommunications see BS 6701 and Appendix I.

Where malfunction may occur (e.g. crosstalk with a communication system) all conductors, including the protective conductor, must be correctly routed and adequate separation provided from other cables (see Appendix I also). A particular form of harmful effect may occur when an electrical installation shares the space occupied by an audio frequency induction loop system — this loop system enables a hearing aid with a telecoil to pick up audio signals from the loop. Under these circumstances, if phase(s) and neutral or switch feeds and switch wires are not close together, 50 Hz interference (and its harmonics) may be picked up by the telecoil of the hearing aid. 528-01

Conventional two-way switching systems can often produce fortuitous 50 Hz radiating loops, but this can be reduced by arranging the circuits as shown in Figure 8.

All the conductors of a circuit should generally follow the same route. Live cables of the same circuit may cause overheating if they enter a ferromagnetic enclosure through different openings. 521-02-01

The EMC Regulations require that attention is paid to electrical equipment and installations to limit their electrical interference from other systems (immunity) and their electrical interference with other systems (emission). 521-02

Electrical power systems will usually have immunity from interference from other systems, but may cause emission interference with sensitive electronic systems. In addition harmonic generation and interference will need to be considered. Electronic systems such as data systems, fire alarms, public address etc. will give the most concern, but detailed information and advice on emission and immunity is not generally available for systems and each installation must be separately considered by experts and any manufacturer's advice acted upon. 515-02

Controlgear and factory-made equipment which is certified or declared to comply with relevant emission and immunity standards can be expected to do so, but on-site a cable installation may pick up or induce interference unless it is adequately separated or protected. The data in Appendix I can be utilised in the absence of other guidance, but no firm data is available to guarantee these figures. The Electrical Contractors' Association (ECA) publication 'Recommended cable separations to achieve electromagnetic compatibility (EMC) in buildings' (reproduced in part in Appendix I) gives some advice which may be useful.

Fig 8: **Circuit for reducing interference with induction loop**

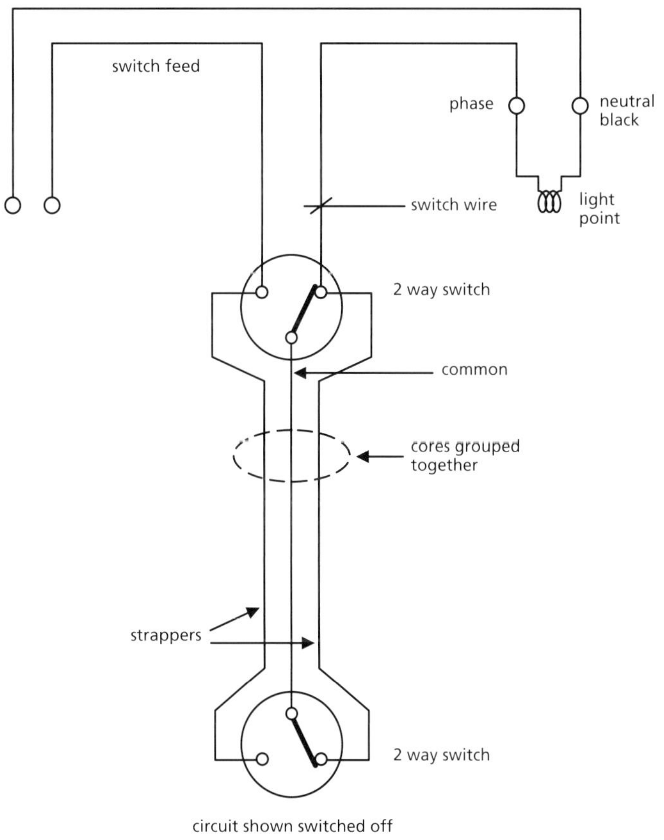

7.5 Placisticiser migration from pvc insulation to the conductor surface

Installers may occasionally encounter a sticky blue/green deposit in equipment or switch and socket-outlet boxes in older installations wired in pvc-insulated cables.

Early manufacture of pvc insulation materials utilised a different formulation from that which is now employed. Thermal cycling of the conductor due to changes in load and conductor temperature drew the thermodynamically unstable insulation plasticiser to the conductor, where, on vertical runs of cable over time the plasticiser liquid worked down the conductor surface to become a sticky blue/green liquid in the accessory box. The material is not corrosive and can be cleaned away with methylated spirit. The blue/green colouring comes from traces of copper absorbed by the liquid. There is no cure for this plasticiser migration, other than replacement of the cable.

As only the plasticiser is removed from the insulation, it is not expected that the insulation resistance of the cable will be affected if the cable is not disturbed but this cannot be guaranteed, and a test should be carried out. The cable condition should also be monitored over time in regular periodic inspections.

Reformulation of pvc compounds in the mid to late 1970's has removed this problem from modern pvc-insulated cables.

Migration also happens when bitumen is in contact with pvc. The bitumen absorbs some plasticiser from the pvc. Whilst the amount of loss is insufficient to have much effect on the properties of the pvc, it is enough to cause considerable reduction in the viscosity of the bitumen, and, the bitumen may become so fluid that it can run.

Section 8 — Installation of Equipment

8.1 Equipment having a high earth leakage current 607

Levels of permitted leakage currents for certain types of equipment are given in Appendix J. The information is taken from relevant British Standards. Where precise information is required the relevant Standards should be consulted.

For appliances incorporating heating elements and motors, the total leakage current is to be within the limits specified in Table J1. Part 3 of BS 3456 (BS EN 60335) covers particular requirements for various types of household appliances, and may alter specific limits from the general requirements of Table J1.

Section 607 of BS 7671 contains supplementary earthing requirements for the installation of equipment having high earth leakage currents. These particular requirements complement the protection described in Section 531, in Chapter 41 and in Chapter 54 of BS 7671. Information technology and other electronic equipment usually incorporates radio frequency filters on the mains conductors. In Class I equipment these filters include capacitors connected between the supply conductors and frame and this results in a standing earth leakage current. 607

On information technology equipment intended for connection to the supply via a 13 A BS 1363 plug, the maximum allowable standing earth leakage current is 3.5 mA as specified in BS EN 60950 : 1992. However modern equipment usually has leakages significantly less than this, usually perhaps 1 mA to 1.5 mA per item.

With desk-top equipment, the use of a number of items of portable equipment connected to a single socket-outlet or a ring or radial final circuit can result in a cumulative leakage current of several milliamperes. This leakage may be high enough to present danger if the protective conductors become open circuit and, also, may operate an RCD. It should be remembered that an RCD to BS 4293, BS EN 61008 or BS EN 61009 is allowed to operate at any value of residual current between 50 % and 100 % of its rated residual operating current. Therefore, unwanted tripping could occur under normal running conditions. This is even more likely if there is a surge of leakage current when switch-on occurs near the peak value of the supply voltage. Regulation 607-02-03 therefore requires that the total earth leakage current should not exceed 25 per cent of the rated residual operating current of an RCD ($I_{\Delta n}$) where this method of protection is utilised. 607-02-03

Information technology equipment connected via BS 4343 (BS EN 60309-2) industrial plugs and socket-outlets is permitted to have higher leakage currents. There is a limit of 10 mA above which the special installation requirements detailed in Section 607 of BS 7671 apply. 607-02-05 607-02-06

These special requirements are aimed at providing a high integrity protective conductor. Should the protective conductor of the circuit feeding the equipment fail, the leakage current could then pass to earth through a person touching the equipment. As with all installations the high integrity protective conductor installation must be periodically inspected and tested to ensure its integrity is maintained (see Guidance Note 3 for further details). See Guidance Note 7 for more detailed information on Section 607 of BS 7671.

8.2 Water heating 554

It is particularly important that manufacturers' instructions are followed when installing electrode water heaters or boilers, to isolate simultaneously every cable feeding an electrode and to provide overcurrent protection. Similarly, the installation of water heaters having immersed and uninsulated heating elements must follow makers' instructions and be correctly bonded. The metal parts of the water heater (e.g. taps, metal-to-metal joints, covers) must be bonded to the incoming water pipe and the pipe connected to the main earthing terminal with a dedicated protective conductor independent of the circuit protective conductor. The heater must be 554-03 554-05 554-05-02 permanently connected through a double-pole switch and not by means of a plug and socket (see Guidance Note 2 also). 554-05-03

8.3 Safety services 56

Safety circuits must have supplies which comply with Chapter 56. The design and construction of any other circuit supplying a safety service must be such that a fault or change in the other circuit does not affect the correct functioning of the safety service. The routes of cables should not render the safety service liable to abnormal fire risk. Such safety services would include: 563-01-01 563-01-02

(i) fire alarms

(ii) emergency lighting

(iii) designated lifts for use in emergency.

8.4 Other equipment 55

Chapter 55 covers requirements for accessories and other equipment, and most of the requirements are reasonably straight forward. Specific requirements are given for typical equipment and accessories. Attention is drawn to Regulation 553-04-01 which allows a choice of accessories to terminate a lighting point. 553 04 01

In larger installations, for convenience of maintenance ıt may be more practicable to provide a luminaire supporting coupler (LSC) or plug and socket-outlet arrangement to facilitate the removal of luminaires for repair or rearrangement of the lighting layout. Item (v) of Regulation 533-04-01 can include any type of socket-outlet for convenience, but the provision of a socket-outlet with a higher current rating than the circuit current rating may require local labelling if it were able to be utilised for other purposes. 553-04-01

The installation of LSCs allows an installation to be completed, tested and energised before luminaires are selected and installed. Luminaires could then be fitted (perhaps after decorating) without interference with the installation, and consequent retesting.

Many lighting installations are now controlled by sophisticated management systems with automatic sensors and software control. There may not be a suitable British Standard for the hardware of such systems, but the wiring and the selection and erection of the equipment and luminaire connections must still comply with BS 7671.

Accessories must comply with the relevant British Standards and care must be exercised to ensure damage does not occur. Accessories must be of a type and IP rating appropriate to their location and should no be installed in locations where they may suffer impact, wet or dampness or corrosion unless they are of a design specifically for the application. Regulation 512-06-01 applies. This regulation applies to socket-outlets near sinks which must be of a type resistant to water ingress if they are likely to be splashed. 512-06-01

Care must be taken to ensure that the conductor operating temperature does not allow the permitted temperature rise of the accessory or equipment to be exceeded (Regulations 512-02-01 and 523-01-01). Thermosetting insulated cables to BS 5467 or BS 7211 operate at a conductor temperature of 90 °C when carrying their rated current, and are not suitable for connection to accessories or devices with a maximum operating temperature of 70 °C unless the cables are derated. (See Appendix F Table F5 for conductor temperature details). 512-02-01 523-01-01

B15 and B22 bayonet lampholders must comply with BS 5042 and Regulation 553-03-03 requires a T_2 temperature rating. This can be overlooked by designers and installers. Edison screw Lampholders should comply with BS EN 60238 : 1996, which details temperature ratings. 553-03-03

The installation of autotransformers and step-up transformers must comply with the requirements of BS 7671. Particular care is needed to guard against danger if there is no neutral feed (e.g. on primaries connected across the phases). Where there is a neutral it must be connected to the common terminal of the transformer. Step-up autotransformers are not permitted on an IT system. Multi-pole linked switches must be used so that all live conductors, including the neutral if any, can be simultaneously disconnected from the supply (see also Guidance Note 2). 555-01

8.5 Luminaires

Luminaires should be manufactured to the requirement of BS 4533, or BS EN 60589, but unfortunately some cheaper or imported types may not be. The installer should be wary of the construction and heat dissipation of these types, and luminaires should be installed strictly in accordance with the manufacturers instructions.

Luminaires are usually installed on non-combustible surfaces, defined in BS EN 60598-1 as "material incapable of supporting combustion" and taken to include metal, plaster and concrete. However sometimes it may be necessary to mount luminaires on materials that may be considered to be flammable (e.g. wood and wood based materials of more than 2 mm thick) and in such cases a classified luminaire to BS EN 60598 marked with the symbol

should be selected. In luminaires marked with this symbol the excessive temperatures which may arise in operation or due to the failure of a component will not cause the mounting surface temperature to exceed 90 °C and thereby cause a risk of fire. Originally this mark only covered luminaires with integral control gear but it now applies to all luminaires complying with BS EN 60598.

Luminaires which cannot comply with this criteria must carry a warning notice or have the following symbol.

For simple pendants and wall lights the installation of the luminaire can have some effect on the compliance and it may be necessary for the designer to specify the minimum length of flexible cord or specific mounting details to limit the heat output onto local surfaces. This is especially significant for high power uplighters. (See Guidance Note 4 for further information on thermal effects and protection.)

For recessed luminaires (e.g. downlighters ELV luminaires etc.) to BS EN 60598 the marking is amended to indicate the temperature rise criteria when in insulated ceilings or enclosed in thermal insulation. The mark is

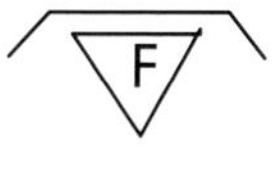

For recessed luminaires not suitable for such installation the mark will be shown crossed out, (as previously).

The fire rating of any ceiling or wall which is penetrated to install recessed luminaires must be maintained after the installation of the luminaires and Regulation 527-01-01 requires the designer to 527-01-01
specify how this is to be carried out. Further information on fire protection requirements can be found in Approved Document B of the 1991 Building Regulations and Guidance Note 4. The luminaire manufacturer's advice and instructions must also be followed.

From a thermal view the temperature of illuminated objects is also important and the heat from a wrongly placed luminaire can initiate combustion. Luminaires to BS EN 60598 can also be marked with a symbol to indicate the minimum distance from a lighted object and a maximum lamp wattage.

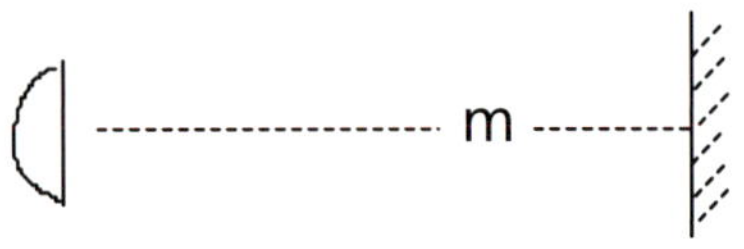

The lamp type is also important and dichroic lamps, where the heat is conducted away through the lamp and luminaire body rather than radiated as part of the light energy, must only be used in the correct types of luminaire. The luminaire manufacturers instructions must be followed in all cases.

Section 526 requires joints between conductors and conductors and equipment to be properly made and enclosed, and Regulation 526-03-02 details suitable 526-03-02
enclosures. This requirement applies to all types of connections in electrical circuits - even to 6 volt and 12 volt ELV luminaires, the connections of which are often below this standard. Regulations 526-03-03 526-03-03
requires the enclosure of the cores of cables from which the sheath has been removed. Not only must the connection between the fixed wiring and the luminaire flexible cable be enclosed but the cores from which the sheath has been removed must also be enclosed. Exposed unsheathed flexible cable is not acceptable on luminaire connections and does not comply with the requirements of BS 7671

8.6 Selection and erection in relation to operation and maintainability 529

BS 7671 Regulation 300-01-01 requires that an assessment be made of the external influences to which the installation is exposed. Amongst these, as indicated in Appendix 5 of BS 7671, is Category BA — dealing with the classification of capability of persons. 529 300-01 Appx 5

A prerequisite of electrical installation work to BS 7671 is that the person undertaking such work is competent to do so. There is no specific definition of competence, but it should be taken to require that the person at least has an understanding of the system to be worked on and its operation, can undertake the necessary installation work and handle the materials properly and safely, and can perform tests adequately and safely and understand the results. Competence also includes a knowledge of safety and safe working practices. Other qualities also may be necessary, depending on the work.

BS 7671 Appendix 5 classification BA recognises different levels from an ordinary person to a skilled person. It must be remembered that a skilled person will not be skilled in all facets of electrical work and consequently will not be competent in all works. BS 7671 uses the terms 'skilled person' and 'instructed person', which are defined in Part 2. Part 2

To require, or allow, a person who is not competent to undertake electrical work may be a breach of the statutory Health and Safety legislation, including the Electricity at Work Regulations 1989.

The Electricity at Work Regulations 1989 specifically place a duty on installation or equipment owners or managers (Duty Holders) to ensure the installation or equipment is utilised safely, and persons are protected from danger. Ordinary persons therefore must not be allowed to gain access to live electrical systems or equipment or carry out electrical work.

All electrical installations should be as simple as possible. All manual components generally accessible should be of the simplest kind. Where not immediately obvious, operating instructions should be displayed adjacent to the component. All other equipment should be located behind lockable doors or accessible only by the use of a tool. The building operator should be provided with full instructions as to the operation and maintenance of the installation, including mechanical operation.

BS 7671 requires an assessment of expected maintenance to be made as part of the design process. Selection of equipment should be relevant to the maintenance (or lack of it) it may receive during its installed and operating life. 341-01-01

Luminaires should be located so that they may be safely relamped and repaired by maintenance personnel from a horizontal surface local to the luminaire. The maintenance personnel should be able to stand directly upon the surface or upon steps at a reasonable height. Where this cannot be achieved consideration should be given to supplementary means of access and/or the luminaire should be equipped with hoisting access or hoisting equipment. 529-01-02

For all items that provide physical protection which may need removal (including lids of cable boxes, trunking, etc.) and replacement, the procedure should be as simple as possible. It should be possible to be effected by one person without assistance. Where frequent access is likely, lids etc. should be hinged. It must be possible to close all lids and covers without putting pressure on enclosed cables and equipment.

Electrical distribution equipment and switchgear should be located so that all components may be operated and maintained safely. The Electricity at Work Regulations 1989 (Regulation 15) specifically require adequate working space, access and lighting to be provided for all electrical equipment, working on which may give rise to danger. 529-01-02

Where appropriate and to prevent danger suitable means shall be provided to isolate equipment and machinery from all sources of supply. This must include any control systems, even at extra-low voltage, as well as power supplies. The isolation devices must be able to be secured against inadvertent operation and clearly labelled if their use is not obvious. 461-01-05

In the case of equipment supplied through a plug and socket-outlet, accessible for withdrawal of the plug, such separation of the plug and socket-outlet is sufficient. For built in domestic appliances (such as washing machines) if the plug and socket-outlet are not accessible unless the machine is moved they may not be suitable for isolation. In any event a plug and socket-outlet must not be selected (specified) as an emergency switching device.

In consumer units and distribution boards provided with multi-terminal neutral bars and earthing bars, the neutral conductors and circuit protective conductors should be connected to their respective bars in the same order as the line conductors are connected to the fuses or circuit-breakers. This will facilitate the disconnection of particular circuits and avoid confusion which might cause accidents.

Live incoming connections within electrical enclosures that cannot be isolated, should be provided with barriers and wherever possible be located in a separate compartment within the main enclosure, and fitted with a warning notice.

There is specific statutory legislation in the form of the Health and Safety at Work Act 1974 etc. and the Electricity at Work Regulations 1989 which include requirements for safe installation, operation and maintenance. Compliance with the requirements of BS 7671 is likely to provide compliance with relevant aspects of this legislation. The Management of Health and Safety at Work Regulations 1992 also provide a statutory duty on employers to analyse health and safety risks in the workplace and this will include risks from the electrical system and electrical equipment.

APPENDICES

CONTENTS

Appendix A: Cable capacities of conduit and trunking

a) General

This Appendix describes a method based on practical work and experimentation which can be used to determine the size of conduit or trunking necessary to accommodate cables of the same size, or differing sizes, and provides a means of compliance with the requirements of Chapter 52 of BS 7671.

The method employs a 'unit system', each cable size being allocated a factor. The sum of all factors for the cables intended to be run in the same enclosure is compared against the factors given for conduit ducting or trunking, as appropriate, in order to determine the size of the conduit or trunking necessary to accommodate those cables.

It has been found necessary, for conduit, to distinguish between:

Case 1, straight runs not exceeding 3 m in length, and

Case 2, straight runs exceeding 3 m, or runs of any length incorporating bends or sets.

The term 'bend' signifies a British Standard 90° bend, and one double set is equivalent to one bend.

For case 1, each conduit size is represented by only one factor. For case 2, each conduit size has a variable factor which is dependent on the length of run and the number of bends or sets. For a particular size of cable the factor allocated to it for case 1 is not the same as for case 2.

For trunking each size of cable has been allocated a factor, as has been each size of trunking.

A number of variables affect any attempt to arrive at a standard method of assessing the capacity of conduit or trunking.

Some of these are:

(i) reasonable care in installation

(ii) acceptable use of the space available

(iii) tolerance in cable sizes

(iv) tolerance in conduit and trunking.

The following tables can only give guidance of the maximum number of cables which should be drawn in. The sizes should ensure an easy pull with low risk of damage to the cables.

Only the ease of drawing-in is taken into account. The electrical effects of grouping are not. As the number of circuits increases the current-carrying capacity of the cable decreases. Cable sizes have to be increased with consequent increase in cost of cable and conduit.

It may therefore be more attractive economically to divide the circuits concerned between two or more enclosures.

The following three cases are dealt with:

Single-core pvc insulated cables to BS 6004 or single-core thermosetting cables to BS 7211

(i) in straight runs of conduit not exceeding 3 m in length. Tables A1 & A2

(ii) in straight runs of conduit exceeding 3 m in length, or in runs of any length incorporating bends or sets. Tables A3 & A4

(iii) in trunking. Tables A5 & A6.

Other sizes and types of cable in conduit or trunking are dealt with in Section e) of this appendix.

For cables and/or conduits, not covered by this Appendix, advice on the number of cables which can be drawn in should be obtained from the manufacturers.

b) Single-core pvc insulated cables in straight runs of conduit not exceeding 3 m in length

For each cable it is intended to use, obtain the factor from Table A1.

Add the cable factors together and compare the total with the conduit factors given in Table A2.

The minimum conduit size is that having a factor equal to or greater than the sum of the cable factors.

TABLE A1
Cable factors for use in conduit in short straight runs

Type of conductor	Conductor cross-sectional area mm^2	Cable factor
Solid	1	22
	1.5	27
	2.5	39
Stranded	1.5	31
	2.5	43
	4	58
	6	88
	10	146
	16	202
	25	385

TABLE A2
Conduit factors for use in short straight runs

Conduit diameter mm	Conduit factor
16	290
20	460
25	800
32	1400
38	1900
50	3500
63	5600

c) Single-core pvc insulated cables; in straight runs of conduit exceeding 3 m in length or in runs of any length incorporating bends or sets

For each cable it is intended to use, obtain the appropriate factor from Table A3.

Add the cable factors together and compare the total with the conduit factors given in Table A4, taking into account the length of run it is intended to use and the number of bends and sets in that run.

The minimum conduit size is that size having a factor equal to or greater than the sum of the cable factors. For the larger sizes of conduit multiplication factors are given relating them to 32 mm diameter conduit.

TABLE A3
Cable factors for use in conduit in long straight runs over 3 m, or runs of any length incorporating bends

Type of conductor	Conductor cross-sectional area mm^2	Cable factor
Solid	1	16
or	1.5	22
stranded	2.5	30
	4	43
	6	58
	10	105
	16	145
	25	217

d) Single-core pvc insulated cables in trunking

For each cable it is intended to use, obtain the appropriate factor from Table A5.

Add all the cable factors so obtained and compare with the factors for trunking in Table A6.

The minimum size of trunking is that size having a factor equal to or greater than the sum of the cable factors.

e) For other sizes and types of cable in conduit or trunking, including flexible conduit

For sizes and types of cable in conduit or trunking other than those given in Tables A1 to A6, the number of cables installed should be such that the resulting space factor does not exceed 35 per cent of the net internal cross-sectional area for conduit and 45 per cent of the net internal cross-sectional area for trunking.

Flexible conduit types may have a smaller internal diameter due to increased wall thickness. The conduit manufacturers' advice should be obtained regarding cable capacity and cable grouping and the required flexibility must be considered. The 35 percent space factor could also be utilised for flexible conduit.

Space factor is defined as the ratio (expressed as a percentage) of the sum of the overall cross-sectional areas of cables (insulation and any sheath) to the internal cross-sectional area of the conduit or other cable enclosure in which they are installed. The effective overall cross-sectional area of a non-circular cable is taken as that of a circle of diameter equal to the major axis of the cable.

The minimum internal radii of bends of cables for fixed wiring as given in Table H1 should be used. Care should be taken to use bends in trunking systems, specifically with larger cables, that allow adequate bending radius.

TABLE A4

Cable factors for runs incorporating bends and long straight runs

length of run m	Conduit diameter, mm																			
	16	20	25	32	16	20	25	32	16	20	25	32	16	20	25	32	16	20	25	32
	Straight				One bend				Two bends				Three bends				Four bends			
1	Covered by Tables A1 and A2				188	303	543	947	177	286	514	900	158	256	463	818	130	213	388	692
1.5					182	294	528	923	167	270	487	857	143	233	422	750	111	182	333	600
2					177	286	514	900	158	256	463	818	130	213	388	692	97	159	292	529
2.5					171	278	500	878	150	244	442	783	120	196	358	643	86	141	260	474
3					167	270	487	857	143	233	422	750	111	182	333	600				
3.5	179	290	521	911	162	263	475	837	136	222	404	720	103	169	311	563				
4	177	286	514	900	158	256	463	818	130	213	388	692	97	159	292	529				
4.5	174	282	507	889	154	250	452	800	125	204	373	667	91	149	275	500				
5	171	278	500	878	150	244	442	783	120	196	358	643	86	141	260	474				
6	167	270	487	857	143	233	422	750	111	182	333	600								
7	162	263	475	837	136	222	404	720	103	169	311	563								
8	158	256	463	818	130	213	388	692	97	159	292	529								
9	154	250	452	800	125	204	373	667	91	149	275	500								
10	150	244	442	783	120	196	358	643	86	141	260	474								

Additional Factors: For 38 mm diameter use . . . . 1.4 x (32 mm factor)
For 50 mm diameter use . . . . 2.6 x (32 mm factor)
For 63 mm diameter use . . . . 4.2 x (32 mm factor)

TABLE A5
Cable factors for trunking

Type of conductor	Conductor cross-sectional area mm^2	PVC, BS 6004 Table 1 Cable factor	Thermosetting BS 7211 Table 3 Cable factor
Solid	1.5	8.0	8.6
	2.5	11.9	11.9
Stranded	1.5	8.6	9.6
	2.5	12.6	13.9
	4	16.6	18.1
	6	21.2	22.9
	10	35.3	36.3
	16	47.8	50.3
	25	73.9	75.4
	35	93.3	95.1
	50	128.7	132.8
	70	167.4	176.7
	95	229.7	227.0
	120	277.6	283.5
	150	343.1	346.4
	185	426.4	433.7
	240	555.7	551.6

Notes:

(i) cable factors are the cable cross-sectional area using the BS upper limit mean overall diameter

(ii) the provision of spare space is advisable, however, any circuits added at a later date must take into account grouping (see Appendix 4 of BS 7671 for further details)

(iii) where thermosetting insulated conductors designed to operate at 90 °C (BS 5467 or BS 7211 etc) are installed together with pvc-insulated conductors designed to operate at 70 °C, it must be ascertained that the pvc-insulated conductors will not be damaged. (Regulation 522-01-01).

TABLE A6
Factors for trunking

Dimensions of trunking mm x mm	Trunking factor	Dimensions of trunking mm x mm	Trunking factor
50 x 38	767	200 x 100	8572
50 x 50	1037	200 x 150	13001
75 x 25	738	200 x 200	17429
75 x 38	1146	225 x 38	3474
75 x 50	1555	225 x 50	4671
75 x 75	2371	225 x 75	7167
100 x 25	993	225 x 100	9662
100 x 38	1542	225 x 150	14652
100 x 50	2091	225 x 200	19643
100 x 75	3189	225 x 225	22138
100 x 100	4252	300 x 38	4648
150 x 38	2999	300 x 50	6251
150 x 50	3091	300 x 75	9590
150 x 75	4743	300 x 100	12929
150 x 100	6394	300 x 150	19607
150 x 150	9697	300 x 200	26285
200 x 38	3082	300 x 225	29624
200 x 50	4145	300 x 300	39428
200 x 75	6359		

Note:

(i) these terms are for metal trunking with trunking thickness taken into account. They may be optimistic for plastic trunking where the cross-sectional area available may be significantly reduced from the nominal by the thickness of the wall material.

f) Background to the tables

The 14th Edition of the IEE Wiring Regulations provided guidance on the number of cables which could be pulled into conduit. Unfortunately the conduit capacities recommended could not always be achieved and guidance with regard to the effect of length, number of bends etc. was only subjective. Further, the tables had been constructed with the use of an arbitrary space factor of 40 % for conduit and 45 % for trunking; this was later shown to be inappropriate.

The criteria adopted for replacement capacity tables were that the length between pulling in points should permit 'easy drawing in' and the cables should not be damaged. These were quite acceptable but, because damage and easy drawing in were not defined, the selection of an appropriate size of conduit for a group of cables remained, to a large extent, dependent on the experience of the installer.

Practical pulling in of cables was carried out over a large range of cable and conduit arrangements to provide a rational means of predicting the size of conduit to accommodate a given bunch of mixed sizes of conduit cables (insulated conductors). The scope was limited to conduit sizes from 16 to 32 mm diameter in both steel and plastic and to single pvc-insulated copper conductors from 1 to 25 mm^2. Single-wire conductors were included for the 1, 1.5 and 2.5 mm^2 sizes and multi-wire conductors were covered for sizes from 1.5 to 25 mm^2. Criteria were that the size of conduit should permit an 'easy pull-in' without insulation damage.

Empirical relationships between numbers and sizes of cables, distance between pull-in points, size of conduit and pulling force were deduced from numerous tests with straight conduit runs using a variety of makes of cable. The effects of type of conduit and ambient temperature were included in the investigation. Considerable variation in results was experienced, but it was possible to determine design centre values for conduit capacities based on a 'unit' approach. The unit system included both solid and stranded conductor cables.

Simple mathematical models were developed during the analysis of the above results which expressed design centre values for conduit capacities using a unit system. For straight conduit runs of 3 m length and upwards and for all lengths which include bends or sets, these models can be used to produce conduit capacity tables or can be adapted for CAD applications. For lengths up to 3 m having no bends or sets a separate empirical table of capacities was developed.

It must be noted that there is no inter-relationship between any of the series of factors for the two different conduit installation cases or for trunking installation, the factors being developed separately for each system.

Appendix B: Degrees of protection provided by enclosures

IP code for ingress protection

General

The requirements of the IP Code are given in BS EN 60529 : 1991. This, however, is a standard used to form the basic requirements of electrical equipment standards, and the construction for IP ratings of a specific type of equipment are given in the BS for that type of equipment (e.g., the requirements for IP ratings and relevant constructional requirements for BS EN 60309-2 (BS 4343) plugs and socket-outlets are given in BS EN 60309). For this reason no mention is made directly to BS EN 60529 in BS 7671.

The degree of protection provided by an enclosure is indicated by two numerals and followed by an optional additional letter and/or optional two supplementary letter(s) in the following way.

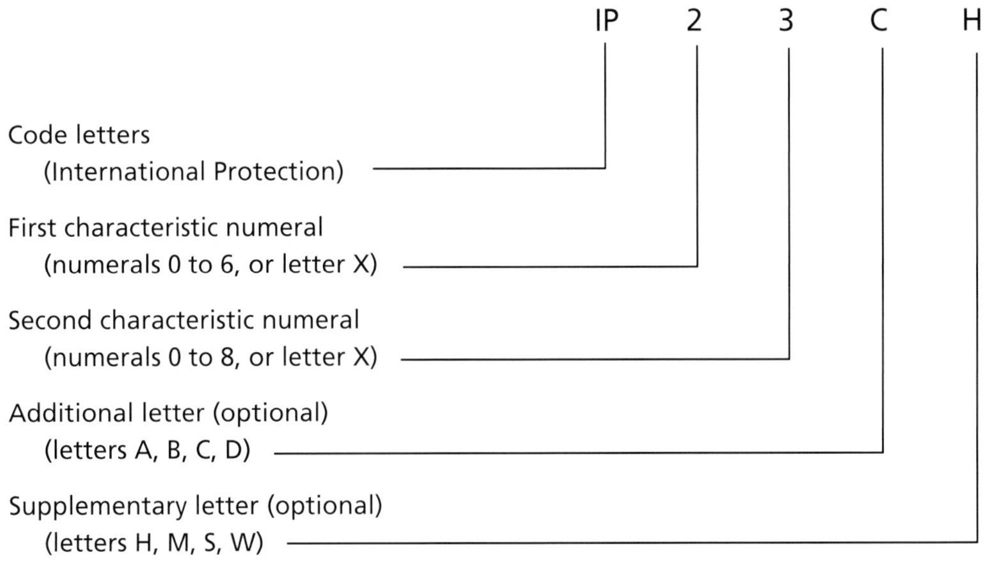

Where a characteristic numeral is not required to be specified, it can be replaced by the letter 'X' ('XX' if both numbers are omitted.

The additional letter and/or supplementary letter(s) may be omitted without replacement. Where more than one supplementary letter is used, the alphabetic sequence shall apply.

Characteristic numerals

First Characteristic Numeral		Second Characteristic Numeral	
(a) Protection of persons against access to hazardous parts inside enclosures (b) Protection of equipment against ingress of solid foreign objects		Protection of equipment against ingress of water	
Number	Degree of Protection	Number	Degree of Protection
0	(a) Not protected (b) Not protected	0	Not protected
1	(a) Protection against access to hazardous parts with the back of the hand (b) Protection against foreign solid objects of 50 mm diameter and greater	1	Protection against vertically falling water drops

Number	Degree of Protection	Number	Degree of Protection
2	(a) Protection against access to hazardous parts with a finger (b) Protection against solid foreign objects of 12.5 mm diameter and greater	2	Protected against vertically falling water drops when enclosure tilted up to 15°. Vertically falling water drops shall have no harmful effect when the enclosure is tilted at any angle up to 15° from the vertical
3	(a) Protection against contact by tools, wires or suchlike more than 2.5 mm thick (b) Protection against solid foreign objects of 2.5 mm diameter and greater	3	Protected against water spraying at an angle up to 60° on either side of the vertical
4	(a) As 3 above but against contact with a wire or strips more than 1.0 mm thick (b) Protection against solid foreign objects of 1.0 mm. diameter and greater	4	Protected against water splashing from any direction

Number	Degree of Protection	Number	Degree of Protection
5	(a) As 4 above (b) Dust-protected (Dust may enter but not in amount sufficient to interfere with satisfactory operation or impair safety)	5	Protected against water jets from any direction
6	(a) As 4 above (b) Dust-tight (No ingress of dust)	6	Protected against powerful water jets from any direction
	No Code	7	Protection against the effects of temporary immersion in water. Ingress of water in quantities causing harmful effects is not possible when enclosure is temporarily immersed in water under standardised conditions
	No Code	8	Protection against the effects of continuous immersion in water under conditions agreed with a manufacturer

Additional letter

The 'additional' letter indicates the degree of protection of persons against access to hazardous parts. It is only used if the protection provided against access to hazardous parts is higher than that indicated by the first characteristic numeral, or if only the protection against access to hazardous parts, and not general ingress, is indicated, the first characteristic numeral being then replaced by an X.

For example, such higher protection may be provided by barriers, suitable shape of openings or clearance distances inside the enclosure.

Refer to BS EN 60529 for full details of the tests and test devices used.

Additional letter	Brief description of protection
A	Protected against access with the back of the hand (minimum 50 mm diameter sphere) (adequate clearance from live parts)
B	Protected against access with a finger (minimum 12 mm diameter test finger, 80 mm long) (adequate clearance from live parts)
C	Protected against access with a tool (minimum 2.5 mm diameter tool, 100 mm long) (adequate clearance from live parts)
D	Protected against access with a wire (minimum 1 mm diameter wire, 100 mm long) (adequate clearance from live parts)

The classification of IPXXB used frequently in BS 7671 indicates that there is no classification for water or dust ingress but protection is provided against access to live parts with a finger.

Supplementary letter

In the relevant product standard, supplementary information may be indicated by a supplementary letter following the second characteristic numeral or the additional letter.

The following letters are currently in use but further letters may be introduced by future product standards.

Letter	Significance
H	High voltage apparatus
M	Tested for harmful effects due to the ingress of water when the movable parts of the equipment (e.g. the rotor of a rotating machine) are in motion
S	Tested for harmful effects due to the ingress of water when the movable parts of the equipment (e.g. the rotor of a rotating machine) are stationary)
W	Suitable for use under specified weather conditions and provided with additional protective features or processes

Product Marking

The requirements for marking a product are specified in the relevant product standard. An example could be:

IP23CS

An enclosure with this designation (IP Code):

(2) protects persons against access to hazardous parts with a finger, and

protects the equipment inside the enclosure against ingress of solid foreign objects having a diameter of 12.5 mm and greater.

(3) protects the equipment inside the enclosure against the harmful effects of water sprayed against the enclosure at an angle of up to 60° either side of the vertical.

(C) protects persons handling tools having a diameter of 2.5 mm and greater and a length not exceeding 100 mm against access to hazardous parts (the tool may penetrate the enclosure up to its full length).

(S) is tested for protection against harmful effects due to the ingress of water when all the parts of the equipment are stationary.

If an enclosure provides different degrees of protection for different intended mounting arrangements, the relevant degrees of protection shall be indicated by the manufacturer in the instructions related to the respective mounting arrangements.

Drip Proof and Splash Proof

Certain electrical equipment has been graded against water ingress by the identifications 'Drip Proof' etc. This labelling is now superseded by the IP coding but is given below with an equivalent IP rating for comparison. Comparisons are not exact, and the designer or installer must satisfy themselves that any equipment is suitable for the location of use.

Drip Proof		IPX2
Rain Proof		IPX3
Splash Proof		IPX4
Jet Proof		IPX5
Protected against Immersion in water		IPX7

IK code for impact protection

General

BS EN 50102 : 1995 'Degrees of protection provided by enclosures for electrical equipment against external mechanical impacts (IK Code)' has introduced a system for classifying the degrees of protection provided by enclosures against mechanical impact. Previously no such classification existed in the UK although some continental countries had their own classifications.

The standard describes only the general requirements and designations for the system — the application of the system to a specific enclosure type will be covered by the British Standard applicable to that equipment or enclosure. An enclosure is defined as a part providing protection of equipment against certain external influences and protection against contact. This may be considered to include conduit, trunking etc.

In general the degree of protection will apply to a complete enclosure. If parts of an enclosure have different degrees of protection, they must be separately identified. The coding is separate from the IP rating and will be marked separately in the following way:

IK 05

Code letters (international mechanical protection) — IK

Characteristic group numeral (00 to 10) — 05

Each characteristic group numeral, represents an impact energy value as shown below:

IK Code	IK00	IK01	IK02	IK03	IK04	IK05	IK06	IK07	IK08	IK09	IK10
Impact energy in Joules	*	0.15	0.2	0.35	0.5	0.7	1	2	5	10	20

* No protection specified

When higher impact energy is required the value of 50 Joules is recommended.

Appendix C: Conventional circuit arrangements and the provision of socket-outlets

a) General

This Appendix gives details of conventional circuit arrangements which satisfy the requirements of Chapter 43 for overload protection and Chapter 46 for isolation and switching, together with the requirements as regards current-carrying capacities of conductors prescribed in Chapter 52 - Selection and erection of wiring systems and Appendix 4 of BS 7671.

It is the responsibility of the designer and installer when adopting these circuit arrangements to take the appropriate measures to comply with the requirements of other Chapters or Sections which are relevant, such as Chapter 41 - Protection against electric shock, Section 434 - Protection against short-circuit current, Chapter 54 - Earthing and protective conductors and the requirements of Chapter 52 other than those concerning current-carrying capacities.

Circuit arrangements other than those detailed in this Appendix are not precluded where they meet the requirements of BS 7671.

The conventional circuit arrangements are:

— final circuits using socket-outlets complying with BS 1363

— final circuits using socket-outlets complying with BS 196

— final radial circuits using socket-outlets complying with BS 4343 (BS EN 60309-2)

— cooker final circuits in household premises

— water heater circuits

— lighting circuits

— electric showers.

b) Final circuits using socket-outlets complying with BS 1363 and fused connection units

Layout

A ring or radial final circuit, with spurs if any, feeds permanently connected equipment and an unlimited number of socket-outlets. Ring final circuits shall comply with the requirements of Regulation 433-02-04 of BS 7671 and the circuit conductors are to form a ring (with spurs as necessary) starting and finishing at the distribution board or consumer unit protective device, neutral bar and, where appropriate, earth bar.

The floor area served by the circuit is determined by the known or estimated load but does not exceed the value given in Table C1.

An assessment of the loading must be made for the design of an installation, in accordance with Chapter 31 of BS 7671, and adequate circuits provided.

For household installations a single 30 A or 32 A ring final circuit may serve a floor area of up to 100 m^2 but consideration should be given to the loading in kitchens which may require a separate circuit. 20 A radial circuits may also adequately serve areas of a dwelling other than kitchens. For other types of premises, final circuits complying with Table E1 may be installed where, owing to diversity, the maximum demand of current-using equipment to be connected is estimated not to exceed the corresponding ratings of the overcurrent protective devices given in that table.

The number of socket-outlets should be such as to ensure compliance with Regulation 553-01-07, each socket-outlet of a twin or multiple socket-outlet unit being regarded as one socket-outlet.

Diversity between socket-outlets and permanently connected equipment has already been taken into account in Table C1 and no further diversity may be applied.

TABLE C1

Typical final circuits using BS 1363 socket-outlets in household installations

Type of circuit	Overcurrent protective device		Minimum conductor size*		Maximum floor area served++
			Copper conductor thermo-setting or pvc insulation	Copper conductor mineral-insulated cables	
	Rating	Type			
	A		mm^2	mm^2	m^2
1	2	3	4	5	6
A1 Ring	30 or 32	Any	2.5	1.5+	100
A2 Radial	30 or 32	Cartridge fuse or circuit-breaker	4	2.5	75
A3 Radial	20	Any	2.5	1.5	50

Notes

* The tabulated values of conductor size may need to be increased where more than two circuits are grouped together, but may be reduced for fused spurs.

Table C1 applies to pvc - or thermosetting-insulated cables of the type identified in Table 4A from Appendix 4 of BS 7671 in installation Method 1 and Method 3, and installed in accordance with those installation methods, and mineral insulated cable (MICC) as identified in Table 4J1A installed in accordance with reference methods 1, 11, 12 and 13 of Table 4A.

\+ 1.5 sq mm two-core MICC installed for a ring final circuit with a protective device rated at not more than 32 A or a radial circuit with a protective device rated at not more than 20 A may be taken as having a conductor not operating at a temperature above 70 °C and may be used to comply with the relevant requirements of Regulation 512-02-01. 1.5 sq mm four core MICC will not comply with Regulation 433-02-04 at 70 °C.

++ The socket-outlets are to be as evenly distributed between circuits as is reasonable.

Where two or more ring final circuits are installed the socket-outlets and permanently connected equipment to be served should be reasonably distributed among the circuits.

Circuit protection

The overcurrent protective device is of the type, and has the rating, given in Table C1.

Conductor size

The minimum size of conductor in the circuit and in non-fused spurs is given in Table C1. However, if the cables of more than two circuits are bunched together or the ambient temperature exceeds 30 °C, the size of conductor is increased and is determined by applying the appropriate correction factors from Appendix 4 of BS 7671 such that the size then corresponds to a current-carrying capacity not less than:

— 20 A for circuit A1 (i.e. 0.67 times the rating of the overcurrent protective device) 433-02-04

— 30 A or 32 A for circuit A2 (i.e. the rating of the overcurrent protective device)

— 20 A for circuit A3 (i.e. the rating of the overcurrent protective device).

The conductor size for a fused spur is determined from the total current demand served by that spur, which is limited to a maximum of 13 A.

When such a spur serves socket-outlets the minimum conductor size is:

— 1.5 mm^2 for pvc or thermosetting insulated cables, with copper conductors

— 1 mm^2 for mineral insulated cables, with copper conductors.

For a 30 A or 32 A ring final circuit with a protective device in accordance with the requirements of Regulation 433-02-04 supplying 13 A socket-outlets to BS 1363, a conductor with a minimum csa of 2.5 sq mm is deemed to comply with the requirements of Regulation 433-02-01.

433-02-04

433-02-01

Electrical accessories to BS 1363 etc. are designed for a maximum operating temperature at their terminals, and are designed for use with pvc-insulated conductors operating at a maximum conductor temperature of 70 °C.

Thermosetting cables can operate with conductor temperatures up to 90 °C and some types of mineral insulated cable can operate with conductor temperatures up to 105 °C.

Such temperatures can damage accessories (see Regulation 522-02-01 of BS 7671). Consequently, when used, such cables must be derated to comply with this requirement (see notes 2 & 5 of Table 4E2A, etc. of Appendix 4 of BS 7671).

Spurs

The total number of fused spurs is unlimited, but the number of non-fused spurs is not to exceed the total number of socket-outlets and items of stationary equipment connected directly in the circuit.

A non-fused spur feeds only one single or one twin socket-outlet or one permanently connected item of electrical equipment. Such a spur is connected to a circuit at the terminals of socket-outlets or at junction boxes or at the origin of the circuit in the distribution board.

A fused spur is connected to the circuit through a fused connection unit, the rating of the fuse in the unit not exceeding that of the cable forming the spur and, in any event, not exceeding 13 A.

Permanently connected equipment

Permanently connected equipment is locally protected by a fuse of rating not exceeding 13 A and controlled by a switch conforming with the requirements of Chapter 46 of BS 7671, or protected by a circuit-breaker of rating not exceeding 16 A.

c) Final radial circuits using 16 A socket-outlets complying with BS 4343 (BS EN 60309-2)

General

A radial circuit feeds equipment the maximum demand of which, having allowed for diversity, is known or estimated not to exceed the rating of the overcurrent protective device and in any event does not exceed 20 A. The number of socket-outlets is unlimited.

Circuit protection

The overcurrent protective device has a rating not exceeding 20 A.

Conductor size

The size of conductor is determined from Appendix 4 of BS 7671, applying the appropriate correction factors and is such that it then corresponds to a current-carrying capacity not less than the rating of the overcurrent protective device.

Types of socket-outlets

Socket-outlets have a rated current of 16 A and are of the type appropriate to the number of phases, circuit voltage and earthing arrangement. Socket-outlets incorporating pilot contacts are not included.

d) Cooker final circuits in household premises

The final circuit supplies a control switch or a cooker control unit (with a socket-outlet). Assessment of the risks associated with the installation of a cooker control unit should always be made. A socket outlet near a cooker could invite accidents if inappropriately located by allowing cables or equipment near hot surfaces. A cooker control unit with a socket-outlet should be installed so as to minimise the risk of such accidents; if this is not possible it is perhaps better that cooker control units should not incorporate a socket-outlet. A modern kitchen should have adequate other socket-outlets.

The rating of a cooking appliance circuit is determined by the assessment of the current demand of the cooking appliance(s), and control unit socket-outlet if any, in accordance with Table H1 of Appendix H. A 30/32 A circuit is suitable for most household cookers, but a 40/45 A circuit may be necessary for larger cookers.

A circuit of rating exceeding 15 A but not exceeding 50 A may supply two or more cooking appliances where these are installed in one room. The control switch should be placed within 2 m of the appliance. Where two stationary cooking appliances are installed in one room, one switch may be used to control both appliances provided that neither appliance is more than 2 m from the switch. Attention is drawn to the need to afford discriminative operation of protective devices as stated in Regulation 533-01-06. 533-01-06

e) Water heating and electric shower final circuits in household premises

The final circuit must supply the full load of the electric shower. No diversity is allowable on the final circuit. Electric showers are usually in the range of 6 kW to 10 kW.

The final circuit supplies a self contained electric shower unit, and should be an individual circuit. A separate local switch may also be installed. Supplementary equipotential bonding is required in accordance with Regulation 601-04-02.

Immersion heaters, fitted to storage vessels in excess of 15 litres capacity, or permanently connected heating appliances forming part of a comprehensive space heating installation should be supplied by their own separate circuits. Immersion heaters should not be connected via a plug and socket, but by a switched flex outlet. However, if an immersion heater was supplied on its own 15 A or 16 A radial circuit, a 15 A plug and socket-outlet could be utilised. Unfortunately such plugs and socket-outlets may not be readily available.

f) Provision of socket-outlets

There is no statutory requirement for the minimum provision of socket-outlets, but the number of socket-outlets needed in a dwelling has risen significantly over the years, and can be expected to continue to rise as more electric and electronic goods become available, and people's working practices change, with working from home becoming more common. Table C2 makes proposals for the provision of socket-outlets in modern dwellings.

553-01-07 Regulation 553-01-07 recommends that adequate socket-outlets should be provided and located to allow the use of portable equipment with only the length of flexible cord normally fitted. Long extension leads, trailing sockets and extending flexible cords which are incorrectly or inappropriately located can create tripping hazards and can damage and cause wear and overheating of the flex when flexible cords are placed under carpets, etc. In order, so far as is possible to make the use of multiway adaptors or extension leads unnecessary an adequate number of fixed socket-outlets should be installed as listed in table C2. Multi-way adaptors long extension leads, trailing socket-outlets and extending flexible cords should never be used unattended and should always be visible and accessible throughout their length.

Socket-outlets must be mounted at a height which will make them accessible but prevent damage by their being struck or wetted by floor cleaning equipment, and allow a comfortable bending radius for larger or stiff flexible cords. (Regulation 553-01-06) 553-01-06
150 mm above finished floor or worktop level is seen as an acceptable minimum but there may be advantage, especially for the elderly or infirm, in mounting socket-outlets at perhaps 1000 to 1200 mm above floor level. The Building Regulations Approved Document M proposes appropriate mounting heights between 450 mm and 1200 mm for electrical equipment for disabled persons. The Approved Document comes into effect on all new and rebuilt buildings and extensions and to new dwellings started on or after 25 October 1999.

Unfortunately, no matter how good overall layout some socket-outlets will inevitably become inaccessible behind furniture when the dwelling is furnished and the designer should consider this in the initial provisions. Also, the continued future growth in appliances and equipment should be expected, and so what may seem initially an over generous provision ultimately may not be. The Copper Development Association (CDA) research into recently built dwellings (under two years old) has shown that generally 50% of house holders considered they did not have an adequate provision of socket-outlets. This rose to 70% in the category of owners of the "normal" three bedroom home.

TABLE C2

Recommended provision of socket-outlets (All socket-outlets are twin)

Location	No. of outlets	Notes
Lounge	6 to 10	(1) (2) (3) (9)
Dining	3	
Kitchen	6 to 10	(3) (4) (5) (9)
Double Bedroom	4 to 6	(3)
Single Bedroom	4 to 6	(3) (6) (9)
Bedsitter	4	
Hall	2	(7)
Stairs/Landing	1	
Loft	1	(7)
Study/Home office	6	(7) (8) (9)
Garage	2	
Utility Room	2	(5)

(This table was prepared with the kind assistance of the ECA, Select, NHBC, CDA and EIEMA).

Notes

(1) The number of outlets depend to some extent on the size of the room, (e.g. through lounge) as larger areas may require more outlets on a pro-rata basis

(2) Two twin socket-outlets should be located close to the TV aerial outlet to allow for TV, video etc. and ancillary equipment supplies

(3) Larger dwellings may require proportionally more socket-outlets pro-rata than smaller dwellings as the occupants of more expensive homes usually have a higher disposable income and hence may have more electrical appliances and equipment.

(4) Kitchens should have the provision of several convenience socket-outlets around work surface areas as well as the provision of specific socket-outlets for washing machines and other appliances where built in. Due to the expected load it may be sensible for the kitchen/utility area socket-outlets to be on a separate radial or ring final circuit.

(5) A lower number of socket-outlets may be appropriate in a kitchen where the washing machine, dryer, freezer etc. are expected to be in a separate utility room.

(6) Bedrooms intended for younger persons should have adequate provision of socket-outlets for computer and electronic equipment.

(7) One twin socket-outlet should be near any telephone outlet for supplies to any telephone, answerphone or fax equipment.

(8) The provisions for an office at home may require more consideration with the user to identify and locate all necessary equipment than is the case with an ordinary domestic installation.

(9) The increasing use of IT and other electrical equipment in the home with high earth leakage currents may require the application of Regulation 607-02-06 to accommodate cumulative leakage currents. 607-02-06

Fig C1: **Heights of switches, sockets etc.**

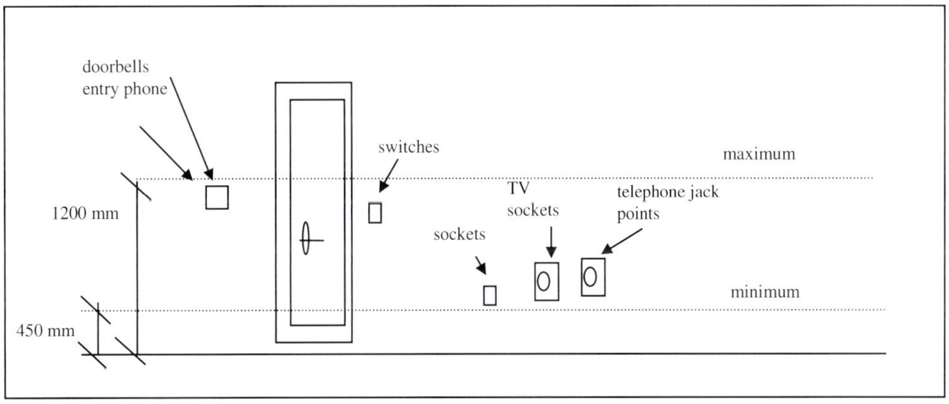

From Approved Document M, 1999 edition Section 8.

Appendix D: Limitation of earth fault loop impedance for compliance with Regulation 543-01-01

Regulation 543-01-01 indicates that the cross-sectional area of a protective conductor, other than an equipotential bonding conductor, shall be: 543-01-01

(i) calculated in accordance with Regulation 543-01-03, or

(ii) determined in accordance with Regulation 543-01-04.

In some cases the type of wiring system it is intended to use determines which of the two methods can be followed. For instance, the widely used flat twin and flat three-core pvc insulated and pvc sheathed cables with protective conductors (cables to Table 5 of BS 6004) do not comply with Table 54G of Regulation 543-01-04 (other than the 1 mm^2 size) and therefore Method (i) must be used.

Where Method (i) is used, in order to apply the formula given in Regulation 543-01-03 it is essential that the time/current characteristic of the overcurrent protective device in the circuit concerned is available. A selection of such characteristics for fuses and miniature circuit-breakers is given in Appendix 3 of BS 7671. For other types of device the advice of the manufacturer has to be sought. The time/current characteristics given in Appendix 3 indicate the values of I_a against various disconnection times for the devices tabulated and give co-ordinates for fixed times. For mcbs the curves indicate the maximum operating current I_a of a particular type of mcb (see Table 3, Section 3.4 for mcb type details). However, for hbc fuses, the curves given in Appendix 3 of BS 7671 are drawn on the median of the 'windows' provided for each rating in the fuse standard.

Assuming that the size and type of cable to be used have already been determined from consideration of other aspects such as mechanical protection and installation requirements together with the magnitude of the design current of the circuit and the limitation of voltage drop under normal load conditions, the first stage is to calculate the earth fault loop impedance, Z_s. If the cable it is intended to use does not incorporate a suitable protective conductor, that conductor has to be chosen separately. Regulation 543-02-02 details what may be used as a protective conductor. 543-02-02

For cables having conductors of cross-sectional area not exceeding 35 mm^2, their inductive reactance can be ignored so that where these cables are used in radial circuits, the earth fault loop impedance Z_s is given by:

$$Z_s = Z_e + (R_1 + R_2) \text{ ohms} \qquad (1)$$

Z_c is that part of the earth fault loop impedance external to the circuit concerned, R_1 is the resistance of the circuit phase conductor from the origin of the circuit where Z_e is taken to the most distant socket-outlet or other point of utilization, and R_2 is the resistance of the circuit protective conductor from the origin of the circuit where Z_e is taken to the most distant socket-outlet or other point of utilization, at normal conductor operating temperature.

Similarly, where such cables are used in a ring circuit without spurs, the earth fault loop impedance Z_s is given by:

$$Z_s = Z_e + 0.25\,R_1 + 0.25\,R_2 \text{ ohms} \qquad (2)$$

Where Z_e is as described in (2) above, R_1 is now the total resistance of the phase conductor between its ends prior to them being connected together to complete the ring, and R_2 is similarly the total resistance of the protective conductor.

Note: Strictly the above equations are vectorial, but arithmetic addition to determine the earth fault loop impedance may be used, as it gives a pessimistically high value for that impedance.

Having determined Z_s, the operating current I (in this case the earth fault current I_f), is given by:

$$I = \frac{U_{oc}}{Z_s} \text{ amperes} \qquad (3)$$

Where:

U_{oc} is the open circuit nominal voltage to earth (phase to neutral voltage) at the origin of the supply, usually the distribution transformer.

For the purposes of this Appendix, in accordance with BS 7671 Appendix 3, the open circuit voltage U_{oc} has been presumed to be 240 V for a nominal supply voltage U_o of 230 V. This allows the formula to become:

$$I_f = \frac{U_o}{Z_s} \qquad (4)$$

Where:

U_o is the nominal phase to earth supply voltage.

From the relevant overcurrent protective device time/current characteristic the time for disconnection (t) corresponding to this earth fault current is obtained.

Substitution for I_f, t and the appropriate k value in the equation

$$S = \frac{\sqrt{I_f^2 t}}{k} \qquad (5)$$

Where:

S is the circuit protective conductor minimum cross-sectional area in mm^2.

from Regulation 543-01-03, this calculation then gives the minimum cross-sectional area of the protective conductor that will provide adequate thermal capacity to withstand the heat produced by the passage of fault current without damage. This has to be equal to or less than the circuit protective conductor size originally chosen (which has resistance R_2). This calculation is usually known as the adiabatic calculation. Where the phase and protective conductors comply with the minimum size requirements of table 54G of BS 7671 this adiabatic calculation is not necessary as the conductors comply with the requirements of Regulation 543-01-04 543-01-04

In order to assist the designer, when the cables it is intended to use are to Table 5 of BS 6004, or Table 7 of BS 7211, or are other pvc-insulated and sheathed cables to the relevant British Standards, Tables D1 to D3 give the maximum earth fault loop impedances (Z_s) for circuits having phase and protective conductors of copper from 1 mm^2 to 16 mm^2 cross-sectional area and where the overcurrent protective device is a fuse to BS 88 Part 2 and Part 6, BS 1361 or BS 3036. The tables also apply if the protective conductor is bare copper and in contact with cable insulated with pvc.

For each type of fuse, two tables are given:

(i) where the circuit concerned feeds socket-outlets and the disconnection time for compliance with Regulation 413-02-09 is 0.4 s, and

(ii) where the circuit concerned feeds fixed equipment only and the disconnection time for compliance with Regulation 413-02-13 is 5 s.

Table D4 gives the maximum earth fault loop impedances for circuits where the overcurrent protective device is a miniature circuit-breaker (mcb) to BS 3871 Part 1 or BE EN 60898. The values given apply to both 0.4 s and 5 s disconnection time since in practice, the overcurrents corresponding to the 0.4 s and 5 s disconnection times both cause the mcb to operate within 0.1 s.

It should always be remembered that an mcb will operate within 0.1 s whenever the fault current is equal to or greater than the upper limit of its 'instantaneous' tripping band. That is:

For Type 1 mcbs	- Fault current ≥ 4 In
For Type 2 mcbs	- Fault current ≥ 7 In
For Type 3 & C mcbs	- Fault current ≥ 10 In
For Type B mcbs	- Fault current ≥ 5 In
For Type D mcbs	- Fault current ≥ 20 In
For Type 4 mcbs	- Fault current ≥ 50 In

Where the adiabatic calculation of S for compliance with Regulation 543-01-03 is carried out, the value of t in equation (5) is 0.1s.

For circuits protected by mcbs, compliance with Table D4 (Table 41B2 of BS 7671) provides compliance with Regulation 543-01-01 where the protective conductors range from 1 mm^2 to 16 mm^2 cross-sectional area and the rated current of the mcbs range from 5 A to 63 A.

In each table the earth fault loop impedances given correspond to the appropriate disconnection time from a comparison of the time/current characteristic of the device concerned and the equation given in Regulation 543-01-03.

The tabulated values apply only when the nominal voltage to earth (U_o) is 230 V.

For guidance on conductor resistances, see Appendix E.

The tabulated loop impedances are design figures assuming final conductor temperatures appropriate to earth fault conditions. For testing purposes the loop impedances must be reduced. For example, if testing at an ambient of 10 °C using cables to Table 5 of BS 6004, the maximum permitted test loop impedance is given by:

$$Z\ test \leq (Z_{FX} - Z_e)\frac{0.96}{1.20} + Z_e \qquad (6)$$

Where:

Z_{FX} is the loop impedance given by Tables D1, D2, D3 or D4.

Z_e is the supply earth loop impedance.

1.20 is the multiplier from Table E2.

0.96 is the ambient correction factor following Table E3.

TABLE D1

Maximum earth fault loop impedance (in ohms) when overcurrent protective device is a fuse to BS 3036

(1) For circuits feeding socket-outlets — 0.4 second disconnection

Protective conductor mm^2	Fuse rating, amperes				
	5	15	20	30	45
1	10	2.67	1.85	NP	NP
1.5	10	2.67	1.85	1.14	NP
2.5 to 16.0	10	2.67	1.85	1.14	0.62

(2) For circuits feeding fixed equipment — 5 second disconnection

Protective conductor mm^2	Fuse rating, amperes				
	5	15	20	30	45
1	18.5	5.58	3.50(ii)	NP	NP
1.5	18.5	5.58	4.00	2.60(ii)	NP
2.5	18.5	5.58	4.00	2.76	1.50(ii)
4 to 16.0	18.5	5.58	4.00	2.76	1.66

Note:

(i) A value for k of 115 from Table 54C of BS 7671 is used for this table. This is also suitable for cables to Table 5 of BS 6004

(ii) Limited by thermal considerations of the protective conductor cross-sectional area, see Regulation 543-01-03.

NP Protective conductor, fuse combination not permitted.

TABLE D2

Maximum earth fault loop impedance (in ohms) when overcurrent protective device is a fuse to BS 88 Part 2 or Part 6

(1) For circuits feeding socket-outlets — 0.4 second disconnection

Protective conductor mm^2	Fuse rating, amperes							
	6	10	16	20	25	32	40	50
1	8.89	5.33	2.82	1.85	1.50	0.86(ii)	NP	NP
1.5	8.89	5.33	2.82	1.85	1.50	1.09	0.83(ii)	NP
2.5 to 16.0	8.89	5.33	2.82	1.85	1.50	1.09	0.86	0.63

(2) For circuits feeding fixed equipment — 5 second disconnection

Protective conductor mm^2	Fuse rating, amperes							
	6	10	16	20	25	32	40	50
1	14.10	7.74	4.00(ii)	2.18(ii)	1.32(ii)	0.84	NP	NP
1.5	14.10	7.74	4.36	3.04	1.63(ii)	1.22	NP	NP
2.5	14.10	7.74	4.36	3.04	2.40	1.92	1.08(ii)	0.71(ii)
4	14.10	7.74	4.36	3.04	2.40	1.92	1.41	1.00(ii)
6 to 16.0	14.10	7.74	4.36	3.04	2.40	1.92	1.41	1.09

Note:

(i) A value for k of 115 from Table 54C of BS 7671 is used for this table. This is also suitable for cables to Table 5 of BS 6004

(ii) Limited by thermal considerations of the protective conductor cross-sectional area, see Regulation 543-01-03.

NP Protective conductor, fuse combination not permitted.

TABLE D3

Maximum earth fault loop impedance (in ohms) when overcurrent protective device is a fuse to BS 1361

(1) For circuits feeding socket-outlets — 0.4 second disconnection

Protective conductor mm²	Fuse rating, amperes				
	5	15	20	30	45
1	10.90	3.43	1.78	1.00(ii)	NP
1.5	10.90	3.43	1.78	1.20	0.43(ii)
2.5 to 16.0	10.90	3.43	1.78	1.20	0.60

(2) For circuits feeding fixed equipment — 5 second disconnection

Protective conductor mm²	Fuse rating, amperes				
	5	15	20	30	45
1	17.10	5.22	2.18(ii)	1.00(ii)	NP
1.5	17.10	5.22	2.80(ii)	1.50(ii)	0.43(ii)
2.5	17.10	5.22	2.93	1.92	0.66(ii)
4	17.10	5.22	2.93	1.92	0.88(ii)
6 to 16.0	17.10	5.22	2.93	1.92	1.00

Note

(i) A value for k of 115 from Table 54C of BS 7671 is used for this table. This is also suitable for cables to Table 5 of BS 6004

(ii) Limited by thermal considerations of the protective conductor cross-sectional area, see Regulation 543-01-03.

NP Protective conductor, fuse combination not permitted.

TABLE D4

Maximum earth fault loop impedance (in ohms) when overcurrent protective device is an mcb to BS 3871 or BS EN 60898

(1) both 0.4 and 5 seconds disconnection times

mcb type	mcb rating, amperes												
	5	6	10	15	16	20	25	30	32	40	45	50	63
1	12.00	10.00	6.00	4.00	3.75	3.00	2.40	2.00	1.88	1.50	1.33	1.20	0.95
2	6.86	5.71	3.43	2.29	2.14	1.71	1.37	1.14	1.07	0.86	0.76	0.69	0.54
B	—	8.00	4.80	—	3.00	2.40	1.92	—	1.50	1.20	1.07	0.96	0.76
3&C	4.80	4.00	2.40	1.60	1.50	1.20	0.96	0.80	0.75	0.60	0.53	0.48	0.38
D	—	2.00	1.20	—	0.75	0.60	0.48	—	0.38	0.30	0.27	0.24	0.19

Selection of a circuit protective conductor

As previously stated Regulation 543-01-01 allows the selection of a circuit protective conductor (cpc) by calculation in accordance with Regulation 543-01-03 or by reference to Table 54G to select the minimum size, in accordance with Regulation 543-01-04.

Table 54G provides a method to establish the minimum size of cpc based on the area of the phase conductor of the circuit. In many cases the material and size of the intended cpc is established by the selection of the wiring system to be used - i.e. steel wire armoured cables, steel conduit, mineral insulated cables etc. If the cross-sectional area of the armour, conduit etc. is found to comply with the requirements of Table 54G then calculation of the thermal capacity is not usually necessary.

Steel conduit and trunking

For some types of cpc, e.g. steel conduit, a check to prove compliance with Table 54G is quite simple, and minimum cross-sectional areas of heavy gauge steel conduit and steel trunkings are listed in Table D5. Taking the value of k_1 as 115 for pvc from Table 43A and k_2 as 47 from Table 54E it can be shown that heavy gauge steel conduit is a suitable protective conductor for all sizes of conductor that can be drawn into it in accordance with Appendix A.

A well installed and maintained steel conduit and cable trunking system can provide an adequate cpc for a final circuit (Regulation 543-02-02). Unfortunately, conduit and trunking impedance data for the calculation of earth fault loop impedances of circuits utilising conduit or trunking as a protective conductor is not readily available and so the practice of installing a separate cpc has arisen. This is usually uneconomical and unnecessary.

However, when a conduit or trunking is to be utilised as a protective conductor, all joints must be tight and secure, and the system protected from corrosion. The cross-section of the protective conductor must be taken as the minimum of any variations of size along the route. Consideration must also be made of the possible further reduction in cross-sectional area at conduit running couplers and cable trunking couplers, with only metal-to-metal contact at bolt holes even when copper links are fitted. Flanged couplers may be necessary to provide adequate continuity at conduit connections onto trunking systems. It must be remembered that the Electricity at Work Regulations 1989, require that an adequate connection to earth must be provided where this is required.

TABLE D5
Cross-sectional areas of steel conduit and trunking

Heavy Gauge Steel Conduit to BS 4568 Pt 1

Nominal diameter (mm)	Minimum steel cross-sectional area (sq mm)
16	53
20	81
25	103
32	135

Steel Surface Trunking BS 4678 Pt 1 (sample sizes)

Nominal size (mm x mm)	Minimum steel cross-sectional area without lid (sq mm)
50 x 50	135
75 x 75	243
100 x 50	216
100 x 100	324
150 x 100	378

Steel Underfloor Trunking BS 4678 Pt 2 (sample sizes)

Nominal size (mm x mm)	Minimum steel cross-sectional area without lid (sq mm)
75 x 25	118
100 x 50	142
100 x 100	213
150 x 100	284

The requirements of Regulation 543-03, as they apply, must be complied with in respect of the preservation of electrical continuity, for a protective conductor common to several circuits. The cross-sectional area must either be calculated to provide adequate capacity for the most onerous conditions of any of the several circuits, or sized from Table 54G to correspond with the largest phase conductor of the several circuits. Therefore a metal conduit or trunking could serve as cpc to all the group of circuits it encloses.

Formula for the calculation of the resistance and inductive reactance values of steel conduit and steel ducting and trunking are published in IEE Guidance Note 6.

Steel-wire armoured cables

If a BS 88 fuse or a miniature circuit breaker is selected for overload protection of PVC/SWA/PVC cable to BS 5346 or XLPE/SWA/PVC cable to BS 5467 then for an earth fault disconnection time of up to 5 s the steel armouring will have a sufficient cross-sectional area to comply with Regulation 643-01-03 with two exceptions. These exceptions are two-core 240 sq mm and 300 sq mm XLPE/SWA/PVC to BS 5467.

From I, t and k, the required minimum cross-sectional area of wire armour required, S, can be calculated and compared with the area provided by the cable. The cross-sectional areas of steel wire armour on PVC insulated cables to BS 6346 and XLPE insulated cables to BS 5467 are given in Tables 8, 9, 10 and 11 of the Commentary on the 15th Edition of the IEE Wiring Regulations.

It will be found that, if the protective device is one of the standard types mentioned in the Regulations and if its rating is not greater than the current rating of the cable, the cross-sectional area of the armour is greater than required to comply with the regulations. This assumes that the earth fault loop impedances of the circuit is such as to ensure that the device will operate in not more than 5 seconds on the occurrence of a fault of negligible impedance.

TABLE D6

Copper conductor, PVC insulated steel wire armoured cables to BS 6346 and fuses to BS 88

Fuse Rating	Required area of SWA (5 second disconnection)	Smallest cable providing required area of SWA			
		2 core	3 core	4 core (equal)	4 core (red. neut.)
A	Sq mm	Sq mm	Sq mm	Sq mm	Sq mm
* 63	14.3	1.5 (20 A)	1.5 (18 A)	1.5 (18 A)	-
80	21	4 (37 A)	4 (31 A)	2.5 (24 A)	-
100	28	10 (69 A)	6 (42 A)	4 (31 A)	-
125	35	10 (69 A)	6 (42 A)	4(31 A)	-
160	46	16 (90 A)	16 (77 A)	10 (58 A)	-
200	61	25 (121 A)	25 (102 A)	16 (77 A)	-
250	77	70 (220 A)	70 (220 A)	50 (155 A)	35 (125 A)
315	107	95 (270 A)	70 (190 A)	50 (155 A)	70 (190 A)
400	133	120 (310 A)	95 (230 A)	70 (190 A)	70 (190 A)
500	178	185 (410 A)	150 (310 A)	120 (310 A)	120 (270 A)
630	224	240 (485 A)	185 (350 A)	150 (310 A)	150 (310 A)
800	350	-	-	400 (550 A)	400 (550 A)

*For BS 88 fuses of 6 A - 50 A ratings, the required armour area is, of course, less than that required for the 63 A fuse and the cables with 1.5 sq mm conductors, the smallest in BS 6346, provide areas exceeding the required minima.

TABLE D7

Copper conductor, PVC insulated steel wire armoured cables to BS 6346 and miniature circuit breakers to BS 3871 and BS EN 60898

(XLPE Cables to BS 5467 also comply with the following requirements)

Circuit Breaker Designation	Required area of SWA (5 second disconnection)	Smallest cable providing required area of SWA		
		2 core	3 core	4 core
	Sq mm	Sq mm	Sq mm	Sq mm
*30 A Type 1	6.1	1.5 (20 A)	1.5 (18 A)	1.5 (18 A)
30 A Type 2 or Type B	10.7	1.5 (20 A)	1.5 (18 A)	1.5 (18 A)
30 A Type 3 or Type C	13.7	1.5 (20 A)	1.5 (18 A)	1.5 (18 A)
50 A Type 1	10.2	1.5 (20 A)	1.5 (18 A)	1.5 (18 A)
50 A Type 2 or Type B	17.8	1.5 (20 A)	1.5 (18 A)	1.5 (18 A)
50 A Type 3 or Type C	21.8	4 (37 A)	4 (31 A)	4 (31 A)

*For 5 A - 20 A circuit breakers of all types to BS 3871 and BS EN 60898 the 1.5 sq mm cables provide armour areas exceeding the required minima.

TABLE D8

Copper conductor, XLPE Insulated steel wire armoured cables to BS 5467 and fuses to BS 88

Fuse Rating	Required area of SWA (5 second disconnection)	Smallest cable providing required area of SWA			
		2 core	3 core	4 core (equal)	4 core (red. neut.)
A	Sq mm	Sq mm	Sq mm	Sq mm	Sq mm
* 25	28.2	16 (114 A)	16 (100 A)	16 (100 A)	-
160	37	16 (114 A)	16 (100 A)	16 (100 A)	-
200	50	35 (190 A)	25 (133 A)	25 (133 A)	25 (133 A)
250	62	50 (228 A)	25 (133 A)	25 (133 A)	25 (133 A)
315	87	95 (356 A)	70 (247 A)	50 (195 A)	50 (195 A)
400	108	95 (356 A)	95 (304 A)	70 (247 A)	70 (247 A)
500	145	185 (540 A)	150 (408 A)	95 (304 A)	95 (304 A)
630	182	185 (540 A)	150 (408 A)	120 (350 A)	150 (408 A)
800	286	-	-	240 (550 A)	300 (627 A)

* For fuses to BS 88 of lower rating than 125 A, cables with 16 sq mm conductors the smallest size in BS 5467, provide armour areas exceeding the required minima.

Tables are also given below identifying which of the standard sizes of steel wire armoured cables can be utilised to comply with Table 54G.

TABLE D9

PVC/SWA/PVC cables to BS 6346. PVC insulation operating at 70 °C.
For Table 54G, k_1 = 115 (103 for 400 sq mm) from Table 43A and k_2 = 51 from Table 54D

sq mm conductor	minimum csa of steel cpc required to comply with Table 54G - sq mm	Actual armour csa from BS 6346		
		2-core	3-core	4-core
1.5	3.4	15	16	17
2.5	5.7	17	19	20
4	9.0	21	23	35
6	13.6	24	36	40
10	22.6	41	44	49
16	36.1	46	50	72
25	36.1	60	66	76
35	36.1	66	74	84
50	56.4	74	84	122
70	79.0	84	119	138
95	107.2	122	138	160
120	135.3	~~(131)~~	150	220
150	169.2	~~(144)~~	211	240
185	208.6	~~(201)~~	230	265
240	270.6	~~(225)~~	~~(260)~~	299
300	338.3	~~(250)~~	~~(289)~~	~~(333)~~
400	403.9	~~(279)~~	~~(319)~~	(467)

~~( )~~ Indicates that the cable does not comply with Table 54G. Therefore the cpc size must be confirmed by calculation as indicated by Regulation 543-01-03 or a supplementary cpc of the full conductor size must be installed.

There are times when an XLPE insulated cable, normally rated at 90 °C, may be run alongside PVC insulated cables operating at 70 °C. Under such circumstances the XLPE insulated cable must be limited to the lower operating temperature of 70 °C. Under these circumstances recourse to Table D10B should be sought to determine if the armour complies with Table 54G.

TABLE D10A

XLPE insulated steel-wire armoured cables to BS 5467 and BS 6724. XLPE insulation operating at 90 °C.

For Table 54G, k_1 = 143 from Table 43A and k_2 = 46 from Table 54D

sq mm conductor	minimum csa of steel cpc required to comply with Table 54G - sq mm	Actual armour csa from BS 5467		
		2-core	3-core	4-core
1.5	4.7	15	16	17
2.5	7.8	17	19	20
4	12.5	19	21	23
6	18.7	22	23	36
10	31.1	~~(26)~~	39	43
16	49.8	~~(41)~~	~~(44)~~	~~(49)~~
25	49.8	~~(42)~~	62	70
35	49.8	62	70	80
50	77.8	~~(68)~~	78	90
70	108.8	~~(80)~~	~~(90)~~	131
95	147.7	~~(113)~~	~~(128)~~	~~(147)~~
120	186.6	~~(125)~~	~~(141)~~	206
150	233.2	~~(138)~~	~~(201)~~	~~(230)~~
185	287.6	~~(191)~~	~~(220)~~	~~(255)~~
240	373.1	~~(215)~~	~~(250)~~	~~(289)~~
300	466.3	~~(235)~~	~~(269)~~	~~(319)~~
400	521.07	~~(265)~~	~~(304)~~	~~(452)~~*

* Not applicable to BS 6724.

~~( )~~ Indicates that the cable does not comply with Table 54G. Therefore the cpc size must be confirmed by calculation as indicated by Regulation 543-01-03 or a supplementary cpc of the full conductor size must be installed.

TABLE D10B

XLPE/SWA cables to BS 5467 and BS 6724, XLPE insulation operating at 70 °C.
For Table 54G, k_1 = 115 from Table 43A and k_2 = 51 from Table 54D

sq mm conductor	minimum csa of steel cpc required to comply with Table 54G - sq mm	Actual armour csa from BS 5467 and BS 6724		
		2-core	3-core	4-core
1.5	3.4	16	17	18
2.5	5.7	17	19	20
4	9.0	19	21	23
6	13.6	22	23	36
10	22.6	26	39	43
16	36.1	41	44	49
25	36.1	42	62	70
35	36.1	62	70	80
50	56.4	68	78	90
70	79.0	80	90	131
95	107.2	113	128	147
120	135.3	~~(125)~~	141	206
150	169.2	~~(138)~~	201	230
185	208.6	~~(191)~~	220	255
240	270.6	~~(215)~~	~~(250)~~	289
300	388.3	~~(235)~~	~~(269)~~	~~(319)~~
400	451.0	~~(265)~~	~~(304)~~	~~(452)~~*

* Not applicable to BS 6724.

~~( )~~ Indicates that the cable does not comply with (Table 54G. Therefore the cpc size must be confirmed by calculation as indicated by Regulation 543-01-03 or a supplementary cpc of the full conductor size must be installed.

Mineral insulated cable with copper sheath

Cross-sectional areas of the copper sheath of light duty and heavy duty mineral insulated cables to BS 6207 are given in the tables below compliance with Table 54G is indicated:

TABLE D11

Cable size reference	Effective sheath area mm^2
500 V Grade (Light duty)	
2L1	5.4
2L1.5	6.3
2L2.5	8.2
2L4	10.7
3L1	6.7
3L1.5	7.8
3L2.5	9.5
4L1	7.7
4L1.5	9.1
4L2.5	11.3
7L1	10.2
7L1.5	11.8
7L2.5	15.4
750 V Grade (Heavy duty)	
1H6.0	8.0
1H10	~~(9.0)~~
1H16	~~(12.0)~~
1H25	~~(15.0)~~
1H35	~~(18.0)~~
1H50	~~(22.0)~~
1H70	~~(27.0)~~
1H95	~~(32.0)~~
1H120	~~(37.0)~~
1H150	~~(44.0)~~
1H185	~~(54.0)~~
1H240	~~(70.0)~~
2H1.5	11
2H2.5	13
2H4	16
2H6	18
2H10	24
2H16	30
2H25	38

TABLE D11 continued

Cable size reference	Effective sheath area mm^2
750 V Grade (Heavy duty)	
3H1.5	12
3H2.5	14
3H4	17
3H6	20
3H10	27
3H16	34
3H25	42
4H1.5	14
4H2.5	16
4H4	20
4H6	24
4H10	30
4H16	39
4H25	49
7H1.5	18
7H2.5	22
12H2.5	34
19H2.5	37

(—) Indicates that the cable does not comply with Table 54G. Therefore the cpc size must be confirmed by calculation as indicated by Regulation 543-01-03 or a supplementary cpc must be installed.

It can be seen that all multicore sizes comply with Table 54G but, with the exception of 1H6 and 1H35 the Heavy duty single-core micc cables do not. Although these cables singularly do not comply, usually single-core cables would be run in pairs for a single-phase circuit, or three or four for a three-phase circuit. The sheath sections, if bonded together, would then sum together and comply for some conductors. (See Regulation 523-05-01 also). 523-05-01

PVC insulated and sheathed cables to BS 6004

These common wiring cables have a cpc of a reduced size from that of the phase conductor for all sizes excepting 1.0 sq mm. Consequently, only the 1.0 sq mm cables (6241Y, 6242Y and 6243Y) will comply with Table 54G. The standard conductor sizes of these cables are identified in Table E1 of Appendix E.

Consequent upon this, calculations are required for the cpc in circuits in which this type of cable is used, usually in domestic installations. The IEE On-Site Guide contains tables giving calculated maximum circuit lengths for various arrangements of conventional circuits and types of protective device. These tables can be used to avoid calculations.

Similar cable size constructions apply to thermosetting cables to Table 7 of BS 7211: 1994. These cables however can have a conductor operating temperature of up to 90 °C. Regulation 523-01-01 requires that equipment and accessories connected to such conductors operating at temperatures exceeding 70 °C must be suitable for operation at such higher temperatures.

Domestic accessories and equipment are not usually suitable for operation at such higher temperatures and consequently cables to BS 7211 cannot be used at their full rating and should be considered as PVC insulated and sheathed types.

Appendix E: Resistance and impedance of copper and aluminium conductors under fault conditions

To check compliance with Regulation 434-03-03 and/or Regulation 543-01-03, i.e. to evaluate the equation $S^2 = I^2 t/k^2$ it is necessary to establish the impedances of the circuit conductors to determine the fault current I and hence the protective device disconnection time t.

Similarly, in order to design circuits for compliance with BS 7671 the limiting values of earth fault loop impedance given in Tables 41B1 and 41B2 and Appendix 3 of BS 7671 or for compliance with the limiting values of the circuit protective conductor Table 41C, it is necessary to establish the relevant impedances of the circuit conductors concerned (see IEE Guidance Note 5 for more information on this subject.)

Where the circuit overcurrent protective device characteristics comply with those given in Appendix 3 of BS 7671, in Regulation 413-02-05 the increase in circuit conductor temperature at fault has been deemed to be taken into account and only the normal conductor operating temperature needs to be considered.

The equation given in Regulations 434-03-03 and 543-01-03 has been based on the assumption of constant value of fault current but in practice that current changes during the period of the fault because, due to the rise in temperature, the conductor resistance increases.

The rigorous method for taking into account the changing character of the fault current is too complicated for practical use and over the range of temperatures encountered in BS 7671 a sufficiently accurate method is to calculate conductor impedances based on the full load current temperature of the conductor (i.e. 70 °C for pvc-insulated conductors, and 90 °C for XLPE-insulated conductors) for devices given in Appendix 3 of BS 7671. This is a change from previous practice, brought in by the first amendment to BS 7671 (Amendment No 1, 1994 (AMD 8536)). It has been recognised that for devices listed in Appendix 3 of BS 7671, the BS testing of these devices includes an allowance for the rise of temperatures during overcurrent tests and taking the final conductor temperature as the average of the initial and final fault temperatures effectively included this allowance twice.

Appendix 3 does not include all types of devices, for example moulded case circuit-breakers (mccbs) to BS EN 60947-2 : 1992. Earth fault loop impedances are not tabulated in BS 7671 for MCCBs nor are characteristics shown in Appendix 3. It may be that these devices do comply with the current Regulation requirements, but there is a wide variety of devices, with adjustable ranges and characteristics, and if there is any doubt, the temperature rise of the average of the initial and final fault temperatures must be utilised. The device manufacturers' advice may be required (see Table E4 also).

The following conductor resistance Table E1 is limited to conductor cross-sectional areas up to and including 35 mm^2 i.e. to conductors having negligible inductive reactance. For larger cables the reactance is not negligible. This reactance is independent of temperature and depends on conductor size and cable make up, and it may be necessary to obtain information from the manufacturer as regards the resistive and reactive components of the impedance of the cables it is intended to use.

Alternatively, values can be calculated for some cables from the volt drop data is given in Appendix 4 of BS 7671. Volt drop data is given in terms of mV/A/m (which is mΩ/m) at full load conductor temperature. Single-phase figures give the resistance/reactance/impedance volt drop for both phase and neutral conductors, which must be divided by two. Three-phase resistance/ reactance/impedance volt drop values are given for a balanced three-phase conductor. The figures must be divided by $\sqrt{3}$ (1.732) to give a figure for the phase conductor only. It should be noted that the reactive component, and consequently the impedance, also varies significantly depending on the cable installation conditions and method. The correct details must be chosen.

Table E1 gives values of ($R_1 + R_2$) per metre for various combinations of sizes of conductors up to and including 35 mm^2 cross-sectional area (see Appendix D for descriptions of R_1 and R_2). It also gives values of resistance per metre for each size of conductor. These values are at given at both 20 °C and 70 °C for PVC insulation when Table 54C applies.

Table E2 gives the multipliers to be applied to the values given in Table E1 for the purpose of calculating the resistance, at maximum permissible operating temperature of the phase conductors and/or circuit protective conductors in order to determine compliance with, as applicable:

(i) earth fault loop impedance of Table 41B1, Table 41B2 or Table 41D

(ii) equation in Regulation 434-03-03

(iii) equation in Regulation 543-01-03

(iv) earth fault loop impedance and resistance of protective conductor of Table 41C

(v) earth fault loop impedances of Appendix D in this Guidance Note.

Where it is known that the actual operating temperature under normal load is less than the maximum permissible value for the type of cable insulation concerned (as given in the tables of current-carrying capacity) the multipliers given in Table E2 may be reduced (see Appendix 4 of BS 7671 for further details.)

TABLE E1

Values of resistance/metre for copper and aluminium conductors and of $(R_1 + R_2)$/metre in milliohms/metre

Cross-sectional area mm^2		Resistance/metre or $(R_1 + R_2)$/metre (mΩ/m) at 20 °C		Resistance/metre or $(R_1 + R_2)$/metre (mΩ/m) at 70 °C in compliance with Table 54C	
Phase conductor	Protective conductor	Plain copper	Aluminium	Plain copper	Aluminium
1	—	18.10		21.72	
*1	1	36.20		43.44	
1.5	—	12.10		14.52	
*1.5	1	30.20		36.24	
1.5	1.5	24.20		29.04	
2.5	—	7.41		8.89	
2.5	1	25.51		30.61	
*2.5	1.5	19.51		23.41	
2.5	2.5	14.82		17.78	
4	—	4.61		5.53	
*4	1.5	16.71		20.05	
4	2.5	12.02		14.42	
4	4	9.22		11.06	
6	—	3.08		3.70	
*6	2.5	10.49		12.59	
6	4	7.69		9.23	
6	6	6.16		7.39	
10	—	1.83		2.20	
*10	4	6.44		7.23	
10	6	4.91		5.08	
10	10	3.66		4.39	
16	—	1.15	1.91	1.38	
*16	6	4.23	—	5.08	
16	10	2.98	—	3.58	
16	16	2.30	3.82	3.58	
25	—	0.727	1.20	0.87	
25	10	2.557	—	3.07	
25	16	1.877	—	2.25	
25	25	1.454	2.40	1.74	
35	—	0.524	0.868	0.63	
35	16	1.674	2.778	2.01	
35	25	1.251	2.068	1.50	
35	35	1.048	1.736	1.26	

* Identifies copper phase/protective conductor arrangement that complies with Table 5 of BS 6004 : 1995 for pvc-insulated and sheathed single, twin or three-core and cpc cables (i.e. 6241Y, 6242Y or 6243Y cables), and similar cable constructions for thermosetting cables to BS 7211 : 1994.

TABLE E2

Operating temperature multipliers to Table E1

Insulation material		70 °C pvc	85 °C rubber	90 °C thermosetting
Multiplier	54B	1.04	1.04	1.04
	54C	1.20	1.26	1.28

The multipliers given in Table E2 are based on the simplified formula given in BS 6360 for both copper and aluminium conductors namely that the resistance temperature coefficient, $\propto$, is 0.004 per °C at 20 °C. This is $R = R_{20}\,[1 + \propto \Theta]$ where Θ is the temperature in degrees Celsius between the initial 20 °C ambient and the final full load conductor temperature, and R_{20} is the conductor resistance at 20 °C.

54B applies where the protective conductor is not incorporated or bunched with cables, or for bare protective conductors in contact with cable covering (assumed initial temperature 30 °C).

54C applies where the protective conductor is a core in a cable or is bunched with cables (assumed initial temperature 70 °C or greater).

Verification

For verification purposes of earth fault loop impedances at installation completion the person carrying out testing will need the maximum values of the phase and protective conductor resistances of a circuit at the ambient temperature expected during tests. This may be different from the reference temperatures of 20 °C used for Table E1.

Table E3 gives correction factors that may be applied to the 20 °C Table E1 values to take account of the ambient temperature (for test purposes only). See IEE Guidance Note 3 - Inspection and Testing - also.

TABLE E3
Ambient temperature multipliers to Table E1

Expected ambient temperature	Correction factor
5 °C	0.94
10 °C	0.96
15 °C	0.98
25 °C	1.02

Tables have been published in theIEE 'On-Site Guide' to provide maximum circuit earth fault loop impedances at 20 °C ready for test comparison (see Appendix 2 of the On-Site Guide).

Devices not listed in Appendix 3

When devices not provided for in Appendix 3 of BS 7671 are used, requiring the application of manufacturers' particular data on the device characteristics, the multipliers in Table E4 are applicable unless decided otherwise by a competent person.

TABLE E4 **Non-standard devices Multipliers to be applied to Table E1 for devices not listed in Appendix 3 for Tables 41B1, 41B2, 41C and 41D**

Insulation material		70 °C pvc*	85 °C rubber	90 °C thermosetting
Multiplier	54B	1.30	1.42	1.48
	54C	1.38	1.53	1.60

* Based on the final temperature of pvc of 160 °C for conductors up to and including 300 sq mm.

These multipliers are based on the limiting fault temperature of conductor insulation. IEC 724:1984 provides the maximum permissible fault temperature limiting values for the following materials:

PVC - to 300 sq mm	160 °C
- over 300 sq mm	140 °C
Butyl (85 °C) Rubber	220 °C
XLPE and EPR	250 °C
Paper	250 °C

Silicone Rubber 350 °C

The formula $R = R_{20} [1 + \propto\Theta]$ is still applicable but the multiplier is higher as the limiting fault temperature is still considered.

Mineral insulated cable with copper sheath

Values of conductor resistance (R_1) and sheath resistance (R_2) per metre for copper sheathed light duty and heavy duty mineral-insulated cables to BS 6207 are given in Table E5.

TABLE E5

Values of R1 and R2 for mineral insulated cable with copper sheath

Cable ref	R_1 Conductor resistance 20 °C	R_2 Sheath resistance 20 °C	R_1 Conductor resistance	R_2 Sheath resistance	R_1 Conductor resistance	R_2 Sheath resistance
			Exposed to touch 70 °C sheath		Not exposed to touch 105 °C sheath	
	Ohms/km	Ohms/km	Ohms/km	Ohms/km	Ohms/km	Ohms/km
500 V Grade (Light duty)						
2L1	18.1	3.95	21.87	4.47	24.5	4.84
2L1.5	12.1	3.35	14.62	3.79	16.38	4.1
2L2.5	7.41	2.53	8.95	2.87	10.03	3.1
2L4	4.81	1.96	5.81	2.22	6.51	2.4
3L1	18.1	3.15	21.87	3.57	24.5	3.86
3L1.5	12.1	2.67	14.62	3.02	16.38	3.27
3L2.5	7.41	2.23	8.95	2.53	10.03	2.73
4L1	18.1	2.71	21.87	3.07	24.5	3.32
4L1.5	12.1	2.33	14.62	2.64	16.38	2.85
4L2.5	7.41	1.85	8.95	2.1	10.03	2.27
7L1	18.1	2.06	21.87	2.33	24.5	2.52
7L1.5	12.1	1.78	14.62	2.02	16.38	2.18
7L2.5	7.41	1.36	8.95	1.54	10.03	1.67

Table E5 continued

Cable ref	R_1 Conductor resistance 20 °C	R_2 Sheath resistance 20 °C	R_1 Conductor resistance	R_2 Sheath resistance	R_1 Conductor resistance	R_2 Sheath resistance
			Exposed to touch 70 °C sheath		Not exposed to touch 105 °C sheath	
	Ohms/km	Ohms/km	Ohms/km	Ohms/km	Ohms/km	Ohms/km
750 V Grade (Heavy duty)						
1H10	1.83	2.23	2.21	2.53	2.48	2.73
1H16	1.16	1.81	1.4	2.05	1.57	2.22
1H25	0.727	1.4	0.878	1.59	0.984	1.72
1H35	0.524	1.17	0.633	1.33	0.709	1.43
1H50	0.387	0.959	0.468	1.09	0.524	1.18
1H70	0.268	0.767	0.324	0.869	0.363	0.94
1H95	0.193	0.646	0.233	0.732	0.261	0.792
1H120	0.153	0.556	0.185	0.63	0.207	0.681
1H150	0.124	0.479	0.15	0.542	0.168	0.587
1H185	0.101	0.412	0.122	0.467	0.137	0.505
1H240	0.0775	0.341	0.0936	0.386	0.105	0.418
2H1.5	12.1	1.9	14.62	2.15	16.38	2.33
2H2.5	7.41	1.63	8.95	1.85	10.03	2
2H4	4.61	1.35	5.57	1.53	6.24	1.65
2H6	3.08	1.13	3.72	1.28	4.17	1.38
2H10	1.83	0.887	2.21	1.005	2.48	1.09
2H16	1.16	0.695	1.4	0.787	1.57	0.852
2H25	0.727	0.546	0.878	0.618	0.984	0.669
3H1.5	12.1	1.75	14.62	1.98	16.38	2.14
3H2.5	7.41	1.47	8.95	1.66	10.03	1.8
3H4	4.61	1.23	5.57	1.39	6.24	1.51
3H6	3.08	1.03	3.72	1.17	4.17	1.26
3H10	1.83	0.783	2.21	0.887	2.48	0.959
3H16	1.16	0.622	1.4	0.704	1.57	0.762
3H25	0.727	0.5	0.878	0.566	0.984	0.613
4H1.5	12.1	1.51	14.62	1.71	16.38	1.85
4H2.5	7.41	1.29	8.95	1.46	10.03	1.58
4H4	4.61	1.04	5.57	1.18	6.24	1.27
4H6	3.08	0.887	3.72	1	4.17	1.09
4H10	1.83	0.69	2.21	0.781	2.48	0.845
4H16	1.16	0.533	1.4	0.604	1.57	0.653
4H25	0.727	0.423	0.878	0.479	0.984	0.518
7H1.5	12.1	1.15	14.62	1.3	16.38	1.41
7H2.5	7.41	0.959	8.95	1.09	10.03	1.18
12H1.5	12.1	0.744	14.62	0.843	16.38	0.912
12H2.5	7.41	0.63	8.95	0.713	10.03	0.772
19H1.5	12.1	0.57	14.62	0.646	16.38	0.698

The calculation of $R_1 + R_2$ for mineral insulated cables is different from the method for other cables, in that the loaded conductor temperature is not usually given. Table 4J of Appendix 4 of BS 7671 gives normal full load sheath operating temperatures of 70 °C for pvc sheathed types and 105 °C for bare cables not in contact with combustible materials, in a 30 °C ambient. Magnesium oxide is a relatively good thermal conductor, and being in a thin layer, it is found that conductor temperatures are usually only some 3 °C higher than sheath temperatures.

However, calculations are complicated as the sheath is of copper which is manufactured to a different material standard to that of the conductors and the $\propto$ factor of 0.004 at 20 °C for normal copper conductor temperature correction does not apply. The $\propto$ value of 0.00275 at 20 °C can be used for sheath resistance change calculations. Table E5 gives calculated values of $R_1 + R_2$ at a standard 20 °C and at standard sheath operating temperatures, and these can be used directly for calculations at full load temperatures.

An estimate of the sheath temperature can be made for a partially loaded cable, in any higher ambient temperature, using the following formula:

Approximate sheath temperature =

$$T_{amb} + \left[\left(\frac{I_b}{I_t} \right)^2 \times 40 \right]$$

Where:

T_{amb} is actual ambient temperature
I_b is circuit design current
I_t is tabulated current-carrying capacity of the cable.

Steel-wire armour, steel conduit and steel trunking

Formulae for the calculation of the resistance and inductive reactance values of the steel-wire armour of cables, steel conduit, ducting and trunking are published in IEE Guidance Note 6.

Generally it is accepted that there is approximately a 10 °C difference between the conductor temperature and the outer sheath temperature for a steel wire armoured cable at full load.

Appendix F: Selection and erection of wiring systems

General

To conform to the requirements of BS 7671, Regulations 511-01-01 and 521-01-01, wiring systems must utilise cables complying with the relevant requirements of the applicable British Standard, or Harmonised Standard. 511-01-01 521-01-01

Alternatively, if equipment complying with a foreign national standard based on an IEC Standard is to be used, the designer or other person responsible for specifying the installation must verify that any differences between that standard and the corresponding British Standard or Harmonised Standard will not result in a lesser degree of safety than that afforded by compliance with the British Standard.

Where equipment to be used is not covered by a British Standard or Harmonised Standard or is used outside the scope of its standard the designer or other person responsible for specifying the installation must satisfy themselves and confirm that the equipment provides the same degree of safety as that afforded by compliance with the Regulations.

A 'wiring system' is defined in Part 2 of BS 7671 as 'an assembly made up of cable or busbars and parts which secure and, if necessary, enclose the cable or busbars'. This can be read to only mean factory made systems, but it is intended to cover all cable types.

Cables are also identified with a voltage grade, to identify the maximum system working voltage for which they are suitable. Conduit wiring cable (6491X) etc. are designated 450/750 V and are harmonised within CENELEC under HD 21 and HD 22. Wiring cables such as twin flat and earth (6242Y) are designated 300/500 V and are not harmonised but are constructed to a British Standard. Armoured cables are designated 600/1000 V and are not harmonised, but are also constructed to a British Standard. There is no difference in utilising types of any of these designations on the UK 230/400 V supply system.

BS 7540 : 1994 is a guide to use for cables with a rated voltage not exceeding 450/750 V and gives installation application advice.

TABLE F1 **Comparison of harmonised cable types to BS 6004**

Type	British Standard Table No	Code Design	Conductor class	Voltage grade	Number of cores	Cross-sectional range mm^2	Temperature °C	
							Installation (minimum)	Storage (maximum)
PVC-insulated, non-sheathed	1	H07V-U	1	450/750	1	1.5 - 16	+5	+40
PVC-insulated, non-sheathed	1	H07V-R	2	450/750	1	1.5 - 630	+5	+40
PVC-insulated, non-sheathed	1	H07V-K	5	450/750	1	1.5 - 240	+5	+40
PVC-insulated, non-sheathed	2	H05V-U	1	300/500	1	0.5 - 1.0	+5	+40
PVC-insulated, PVC-sheathed	3		1/2	300/500	2 - 5	1.5 - 35	+5	+40
PVC-insulated, PVC-sheathed	4	6181Y 6192Y 6193Y	1/2	300/500	1 - 3	1 - 35	+5	+40
PVC-insulated, PVC-sheathed	5	6242Y 6243Y	1/2	300/500	2 - 3	1.0 - 16	+5	+40
PVC-insulated, PVC-sheathed	6	6192Y 6242Y	2	300/500	2	1.5 - 2.5	+5	+40

TABLE F2
Designation system for cables complying with the European Harmonisation Standard

Cable Reference

1 2 3 4 5 - 6 7 8 9

1. Basic Standards

H	Harmonised Standards
A	Authorised National Standards (derived from a harmonised cable standard)
N	Non Authorised National Standards

2. Rated Voltage

03	300/300 Volt
05	300/500 Volt
07	450/750 Volt
1	600/1000 Volt

3 & 4 Insulation & Sheathing Material

B	Ethylene Propylene Rubber (EPR)
E	Polyethylene (PE) (low density LDPE)
E2	Polyethylene (High density) HDPE
E4	Polytetrafluoroethylene (PTFE)
E6	Ethylene Tetrafluoroethylene (ETFE)
E7	Polypropylene (PP)
G	Ethylene-Vinyl-Acetate (EVA)
J	Glass fibre braid (GFB)
N	Polychloroprene (PCP)
N4	Chlorosulphonated Polyethylene (CSP)
Q	Polyurethane (PU)
Q2	Polyethylene-Terephtalate (PETP)
Q4	Polyamide (PA)
R	Natural Rubber
S	Silicone
T	Textile braid
V	Polyvinyl-Chloride (PVC)
V2	Heat Resistant Polyvinyl-Chloride (HR PVC)
X	Cross Lined Polyethylene (XLPE)

5. Special Construction & Shapes

H	Flat Construction with Divisible Cores
H2	Flat Construction, Non Divisible Core
H5	Two or more cores twisted together non sheathed

6. Type of Conductor

A	Aluminium
	Copper (no code letter)
F	Flexible for moveable installations (Class 5 IEC 228)
H	Highly flexible for moveable installations (Class 6 IEC 228)
K	Flexible for fixed installations (Class 5 IEC 228)
R	Stranded (Class 2 IEC 228)
U	Solid (Class 1 IEC 228)
Y	Tinsel

7. Number of Cores

8. Protective Conductor

X	Without protective core
G	With protective core

9. Nominal Cross Section of Conductors in sq mm

Additional Designations

Concentric Conductors & Screens

A	Concentric Aluminium Conductor
C	Concentric Copper Conductor
A7	Aluminium/Laminate Screen
C4	Overall Copper Braid Screen
C5	Cores Individually Copper Braid Screen
C7	Lapped Copper (Wire, Tape or Strip) Screen

Special Components

D3	Central Strainer (Textile or Metallic)
D5	Central Filler (Not Load Bearing)

Armours

Z2	Steel Wire Armour
Z3	Flat Steel Wire Armour
Z4	Steel Tape Armour
Z5	Steel Wire Braid

Reproduced with kind permission of Anixter (UK) Ltd.

TABLE F3

Commonly used cable type reference numbers

* = number of cores (e.g. 2183Y is 3 core)

0361	Welding cable, tinned copper conductor, vulcanised rubber insulation/tough rubber sheath. BS 638.	658*TQ	Circular, EPR insulation/CSP bedding/galv. mild steel wire braid armour/CSP sheathed 600/1000V BS 6883, ship wiring cable (circular conductors).
2093Y	3 core, PVC heat resisting insulation/PVC BS. heat resisting sheathed. Circular, 300/300V 6500, flexible cord	658*TQ(S)	As 658*TQ but with shaped conductors.
218*Y	Circular, PVC insulation/PVC sheathed. 300/300V BS 6500, flexible cord.	694*X	Circular, PVC insulation/PVC bedded/single steel wire armoured/PVC sheathed 600/1000V BS 6346, mains wiring cable.
2192Y	2 core, PVC insulation/PVC sheathed Flat 300/300V. BS 6500, flexible cord.	694*X(S)	As 694*X but with shaped conductors.
2213	3 core, vulcanised rubber insulation/tough rubber sheathed/cotton braid, 300/300V BS 6500, flexible cord.	HO7V-R	Single core, PVC insulation 4 sq. mm - 400 sq. mm, 450/740V to BS 6004.
		HO7V-U	Single core, PVC insulation, 1.5 sq. mm - 2.5 sq. mm, (solid) 450.750V to BS 6004.
318*	Circular, vulcanised rubber insulation/tough rubber sheathed, 300/500V BS 6500, flexible cord.	HO5V-U	Single core, PVC insulation, 1.0 sq mm (solid) 300/400V to BS 6004
318*Y	Circular, PVC insulation/PVC sheathed 300/500V BS 6500, flexible cord.	HO5RR-F	Circular, vulcanised rubber insulation/tough rubber sheathed. 300/500V to BS 6500
318*P	Circular, vulcanised rubber insulation/PCP sheathed, 300/500V BS 6500, flexible cord.	HO3VV-F	Circular, PVC insulated/PVC sheathed, 300/300V to BS 6500.
318*TQ	Circular, EPR insulation/CSP sheathed 300/500V BS 6500, flexible cord.	HO3VVH2-F	Flat 2 core, PVC insulated/PVC sheathed, 300/300V to BS 6500.
618*Y	Circular, PVC insulation/PVC sheathed 1 sq mm - 35 sq mm 300/500V BS 6004 50 sq mm - 630 sq mm gen in acc BS 6346, wiring cable.	HO5VV-F	Circular, PVC insulated/PVC sheathed, 300/500V to BS 6500.
		HO5VVH2-F	Flat, 2 core, PVC insulated/PVC sheathed, 300/500V to BS 6500.
619*Y	Flat, PVC insulation/PVC sheathed 300/500V BS 6004, wiring cable.	HO5V-K	Single core, PVC insulation 0.5 sq. mm - 1.0 sq. mm 300/500V to BS 6500.
624*Y	Flat, PVC insulation - laid up in parallel with a bare CPC within one interstice/PVC sheathed 300/500V BS 6004, wiring cable.	HO7V-K	Single core, PVC insulation, 1.5 sq. mm - 2.5 sq. mm, 450/750V to BS 6004.
6491X	Single core, PVC insulation 1 sq. mm - 630 sq. mm. 450/750V BS 6004.	HO7RN-F	Circular, rubber insulated/rubber sheathed, 450/750V to BS 6007.
657*TQ	Circular, EPR insulation/CSP sheathed 600/1000V BS 6883, ship-wiring cable. (circular conductors).	(AL)	To denote aluminium conductors. To be added after other suffix, if used e.g. 6491X(AL) and 6242Y(AL).
657*TQ(S)	As 657*TQ but with shaped conductors	(S)	Shaped conductors.

Reproduced with kind permission of Anixter (UK) Ltd.

TABLE F4
Applications of cables for fixed wiring

Type of cable	Uses	Comments
PVC, thermosetting or rubber insulated non-sheathed	In conduits, cable ducting or trunking	(i) intermediate support may be required on long vertical runs (ii) 70 °C maximum conductor temperature for normal wiring grades — including thermosetting types (4) (iii) cables run in pvc conduit shall not operate with a conductor temperature greater than 70 °C (4)
Flat pvc or thermosetting insulated and sheathed	(i) general indoor use in dry or damp locations. Maybe embedded in plaster (ii) on exterior surface walls, boundary walls and the like (iii) overhead wiring between buildings (6) (iv) underground in conduits or pipes (v) in building voids or ducts formed in situ	(i) additional protection may be necessary where exposed to mechanical stresses (ii) protection from direct sunlight may be necessary. Black sheath colour is better for cables in sunlight (iii) see Note (4) (iv) unsuitable for embedding directly in concrete (v) may need to be hard drawn (HD) copper conductors for overhead wiring (Note 6)
Split-concentric pvc insulated and sheathed	General	(i) additional protection may be necessary where exposed to mechanical stresses (ii) protection from direct sunlight may be necessary. Black sheath colour is better for cables in sunlight
Mineral insulated	General	With overall pvc covering where exposed to the weather or risk of corrosion, or where installed underground, or in concrete ducts
PVC or XLPE insulated, armoured, pvc sheathed	General	(i) additional protection may be necessary where exposed to mechanical stresses (ii) protection from direct sunlight may be necessary. Black sheath colour is better for cables in sunlight
Paper insulated, lead sheathed and served	General, for main distribution cables	With armouring where exposed to severe mechanical stresses or where installed underground

Notes:

(1) The use of cable covers (preferably conforming to BS 2484) or equivalent mechanical protection is desirable for all underground cables which might otherwise subsequently be disturbed. Route marker tape should also be installed, buried just below ground level.

(2) Cables having pvc insulation or sheath should preferably not be used where the ambient temperature is below 0 °C, or has been within the preceding 24 hours. Where they are to be installed during a period of low temperature, precautions should be taken to avoid risk of mechanical damage during handling. A minimum ambient temperature of 5 °C is advised in BS 7540 : 1994 for some types of pvc insulated and sheathed cables.

(3) Cables must be suitable for the maximum ambient temperature, and shall be protected from any excess heat produced by other equipment, including other cables.

(4) Thermosetting cable types (to BS 7211 or BS 5467) can operate with a conductor temperature of 90 °C. This must be limited to 70 °C when drawn into a conduit etc. with pvc insulated conductor (521-07-03) or connected to electrical equipment (512-02-01 and 523-01-01), or when such cables are installed in plastic conduit or trunking.

(5) For cables to BS 6004, BS 6007, BS 7211, BS 6346, BS 5467 and BS 6724, further guidance may be obtained from those standards. Additional advice is given in BS 7540 :1994 'Guide to use of cables with a rated voltage not exceeding 450/750 V' for cables to BS 6004, BS 6007 and BS 7211.

(6) Cables for overhead wiring between buildings must be able to support their self-weight and any imposed wind or ice/snow loading. A catenary support is usual but hard drawn copper types may be used.

TABLE F5

Applications of flexible cables and cords to BS 6007 : 1993, BS 6141 : 1991 and BS 6500 : 1994 generally

Type of flexible cord	Uses
Light pvc insulated and sheathed	Indoors in household or commercial premises in dry situations, for light duty
Ordinary pvc insulated and sheathed	(i) indoors in household or commercial premises, including damp situations, for medium duty (ii) for cooking and heating appliances where not in contact with hot parts (iii) for outdoor use other than in agricultural or industrial applications (iv) electrically powered hand tools
60 °C rubber insulated braided twin and three-core	Indoors in household or commercial premises where subject only to low mechanical stresses
60 °C rubber insulated and sheathed	(i) indoors in household or commercial premises where subject only to low mechanical stresses (ii) occasional use outdoors for supplies to equipment (iii) electrically powered hand tools
60 °C rubber insulated oil-resisting and flame-retardant sheath	(i) general, unless subject to severe mechanical stresses (ii) fixed installations protected in conduit or other enclosures
85 °C rubber insulated HOFR sheathed	General, including hot situations e.g. night storage heaters and immersion heaters
85 °C heat resisting pvc insulated and sheathed	General, including hot situations e.g. for pendant luminaires
150 °C rubber insulated and braided	(i) at high ambient temperatures (ii) in or on luminaires
185 °C glass-fibre insulated single-core twisted twin and three-core	For internal wiring of luminaires only and then only where permitted by BS 4533
185 °C glass-fibre insulated braided circular	(i) dry situations at high ambient and not subject to abrasion or undue flexing (ii) wiring of luminaires

Notes:

(1) Cables and cords having pvc insulation or sheath should preferably not be used where the ambient temperature is consistently below 0 °C. Where they are to be installed during a period of low temperature, precautions should be taken to avoid risk of mechanical damage during handling.

(2) Cables and cords shall be suitable for the maximum ambient temperature, and shall be protected from any excess heat produced by other equipment, including other cables.

(3) For flexible cords and cables to BS 6007, BS 6141 and BS 6500 further guidance may be obtained from those standards, or from BS 7540 : 1994 'Guide to use of cables with a rated voltage not exceeding 450/750 V'.

(4) When used as connections to equipment flexible cables and cords should be of the minimum practical length to minimise danger and in any case of such a length that allows the protective device to operate correctly.

(5) When attached to equipment flexible cables and cords should be protected against tension, crushing, abrasion, torsion and kinking particularly at the inlet point to the electrical equipment. At such inlet points it may be necessary to use a device which ensures that the cable is not bent to too an internal radius below that given in the appropriate part of Table 4 of BS 6700. Strain relief, clamping devices or cord guards should not damage the cord.

(6) Flexible cables and cords should not be used under carpets or to other coverings, or where furniture or other equipment may rest on them. Flexible cables and cords should not be placed where there is a risk of damage from traffic passing over them.

(7) Flexible cables and cords should not be used in contact with or close to heated surfaces especially if the surface approaches the upper thermal limit of the cable or cord.

British Standards

Most cables in general use today are manufactured to a specific British Standard (or Standards) that covers the design, materials, manufacture, testing and performance of the cable. The designer of the installation has to select a cable type that is suitable for the performance required and the relevant British Standard is usually specified. 511-01 522 Appx 5

The cable manufacturer will provide full details of cables and the standards to which they are manufactured, and BS 7671 Appendix 4 indicates the relevant British Standards for each cable type on the current rating tables. A full listing of British Standards referred to in BS 7671 is provided in Appendix 1.

Fire stopping

All cable routes that pass between building fire zones or areas, must be adequately sealed against the transmission of flames and/or smoke between zones or areas. 527-02

The specification for such fire stopping is outside the scope of the Guidance Note but should provide fire resistance to a standard at least equal to the original element of the building construction. The specification for fire stopping should be detailed by the Project Architect or Designer for all services, (i.e. cables, pipes etc.) and should comply with the requirements of the Building Regulations, as applicable. Intumescent materials are commonly used.

All voids within ducts, trunking etc should be filled, as well as the space around such ducts and trunkings where they pass through walls etc. It is not usual however to seal the inside of conduits, except in classified hazardous areas, as required by the relevant codes or standards. 527-02-02

Appendix G: Notes on methods of support for cables, conductors and wiring systems

This Appendix described methods of support for cables, conductors and wiring systems which satisfy the relevant requirements of Chapter 52 of BS 7671. The use of other methods is not precluded where specified by a competent person, having due regard for protection of the cable from mechanical damage. The methods described in this Appendix make no specific provisions for fire or thermal protection of the cables. All cable installations and routings must be carefully considered and protection provided as required. The advice of the local authority and the Fire Officer must be taken as necessary on the project. The Project Architect and Designer must also comply with the CDM Regulations (see IEE Guidance Note 4 also).

Cables generally

Items (i) to (ix) below are generally applicable to supports on structures which are subject only to vibration of low severity and a low risk of mechanical impact:

(i) for non-sheathed cables, installation in conduit without further fixing of the cables, precautions being taken against undue compression or other mechanical stressing of the insulation at the top of any vertical runs exceeding 5 m in length

(ii) for cables of any type, installation in factory made ducting or trunking without further fixing of the cables, vertical runs not exceeding 5 m in length without intermediate support

(iii) for sheathed and/or armoured cables drawn into ducts formed in situ in the building structure. The internal surfaces of the duct must be protected to prevent abrasion of the cables, especially when drawing in. Ducts should also be adequately sealed against the spread of fire and smoke. Vertical runs not exceeding 5 m in length without intermediate support

(iv) for sheathed and/or armoured cables installed in accessible positions, support by clips at spacings not exceeding the appropriate value stated in Table G2

(v) for cables of any type, resting without fixing in horizontal runs of ducts, conduits, factory made cable ducting or trunking

(vi) for sheathed and/or armoured cables in horizontal runs which are inaccessible and unlikely to be disturbed, resting without fixing on part of a building, the surface of that part being reasonably smooth

(vii) for sheathed and/or armoured cables in vertical runs which are inaccessible and unlikely to be disturbed, supported at the top of the run by a clip and a rounded support of a radius not less than the appropriate value stated in Table G1

(viii) for sheathed cables without armour in vertical runs which are inaccessible and unlikely to be disturbed, supported by the method described in (vii) above; the length of run without intermediate support not exceeding 2 m for a lead sheathed cable or 5 m for a rubber or pvc sheathed cable

(ix) for rubber or pvc sheathed cables, installation in conduit without further fixing of the cables, any vertical runs being in conduit of suitable size and not exceeding 5 m in length.

Cables in particular conditions

(x) for flexible cords used as pendants, attachment to a ceiling rose or similar accessory and lampholder by the cord grip or other method of strain relief provided in the accessories (see Table 4H3A of BS 7671 for maximum mass supportability flexible cards)

(xi) for temporary installations and installations on construction sites, protection and supports so arranged that there is no appreciable mechanical strain on any cable termination or joint, and the cables are protected from mechanical damage

(xii) in caravans, for sheathed flexible cables or cords in inaccessible spaces such as ceiling, wall and floor spaces, support at intervals not exceeding 0.25 m for horizontal runs and 0.4 m for vertical runs. In caravans for horizontal runs of sheathed flexible cables or cords passing through floor or ceiling joists in inaccessible floor or ceiling spaces, securely bedded in thermal insulating material, no further fixing is required. (See Regulation 608-06-01 for details of caravan wiring systems.)

Overhead wiring

(xiii) for cables sheathed with rubber or pvc, supported by a separate catenary wire, either continuously bound up with the cable or attached thereto at intervals not exceeding those stated in Column 2 of Table G2; and the minimum height above ground being in accordance with Table G3

(xiv) support by a catenary wire incorporated in the cable during manufacture, the spacings between supports not exceeding those stated by the manufacturer and the minimum height above ground being in accordance with Table G3

(xv) for spans without intermediate support (e.g. between buildings) of multicore sheathed cable, with or without armour, terminal supports so arranged that no undue strain is placed upon the conductors or insulation of the cable, adequate precautions being taken against any risk of chafing of the cable sheath, and the minimum height above ground and the length of such spans being in accordance with the appropriate values indicated in Table G3. Hard drawn (HD) copper conductors may be necessary for longer spans. Cable manufacturers' advice should be complied with

(xvi) bare or insulated conductors of an overhead line for distribution between a building and a remote point of utilisation (e.g. another building) supported on insulators, the lengths of span and heights above ground having the appropriate values indicated in Table G3 or otherwise installed in accordance with the Electricity Supply Regulations 1988 (as amended)

(xvii) for spans without intermediate support (e.g. between buildings) and which are in situations inaccessible to vehicular traffic, insulated or multicore sheathed cables installed in heavy galvanised steel conduit, the length of span and height above ground being in accordance with Table G3, provided that the conduit shall be earthed in accordance with Part 4 and 5 of BS 7671, shall be securely fixed at the ends of the span, and shall not be jointed in its span.

Conduit and cable trunking

(xviii) rigid conduit supported in accordance with Table G4

(xix) cable trunking supported in accordance with Table G5

(xx) conduit embedded in the material of the building, suitably treated against corrosion if necessary

(xxi) pliable conduit embedded in the material of the building or in the ground, or supported in accordance with Table G4.

TABLE G1
Minimum internal radii of bends in cables for fixed wiring

Insulation	Finish	Overall diameter*	Factor to be applied to overall diameter of cable to determine minimum internal radius of bend
XLPE, pvc or rubber (circular, or circular stranded copper or aluminium conductors)	Non-armoured	Not exceeding 10 mm	3(2)†
		Exceeding 10 mm but not exceeding 25 mm	4(3)†
		exceeding 25 mm	6
	Armoured	Any	6
XLPE, pvc or rubber (solid aluminium or shaped copper conductors)	Armoured or non-armoured	Any	8
Mineral insulated, copper sheathed	Bare or pvc covered	Any	6(3)‡
Flexible cables	Sheathed	Any	No specific provision, but no tighter than the equivalent sized non-armoured cable**

Notes:

* For flat cables the diameter refers to the major axis.

**Flexible cables can be damaged by too tight, or repeated bending.

† The figure in brackets relates to single-core circular conductors of stranded construction installed in conduit, ducting or trunking.

‡ For mineral insulated cables, the bending radius should normally be limited to a minimum of 6 times the diameter of the bare copper sheath, as this will allow further straightening and re-working if necessary. However, cables may be bent to a radius not less than 3 times the cable diameter over the copper sheath, provided that the bend is not re-worked.

Conduit bends

The radius of every conduit bend must be such as to allow compliance with the minimum bending radii of cables installed in the conduit (e.g. see table above). In addition the inner radius of the bend shall not be less than 2.5 times the outside diameter of the conduit.

TABLE G2

Spacings of supports for cables in accessible positions the entire support derived from the clips

		Maximum spacings of clips							
		Non armoured rubber, plastics, or lead sheathed cables							
		Generally		In caravans		Armoured cables		Mineral insulated copper sheathed cables	
Overall (1) diameter of cable		(2) Horizontal	(2) Vertical	(2) Horizontal	(2) Vertical	(2) Horizontal	(2) Vertical	(2) Horizontal	(2) Vertical
	1	2	3	4	5	6	7	8	9
	mm	mm	mm	mm	mm	mm	mm	mm	mm
	≤9	250	400			—	—	600	800
>9	≤15	300	400			350	450	900	1200
>15	≤20	350	450	250 (for all sizes)	400 (for all sizes)	400	550	1500	2000
>20	≤40	400	550			450	600	2000	3000

Notes:

For the spacing of supports for cables having an overall diameter exceeding 40 mm, or conductors of cross-sectional area 300 mm^2 and larger, the manufacturer's recommendations should be observed.

(1) For flat cables taken as the dimension of the major axis.

(2) The spacings stated for horizontal runs may be applied also to runs at an angle of more than 30° from the vertical. For runs at an angle of 30° or less from the vertical, the vertical spacings are applicable.

(3) The spacings given in this table are maxima, and for good workmanship in certain circumstances the spacings may need to be reduced.

TABLE G3
Maximum lengths of span and minimum heights above ground for overhead wiring between buildings etc.

Type of system 1	Maximum length of span 2	Minimum height of span above ground		
		At road crossings 3	In positions accessible to vehicular traffic, other than crossings 4	In positions inaccessible to vehicular traffic (†) 5
	m	m	m	m
Cables sheathed with plastics or having an oil-resisting and flame-retardant (hofr) sheath without intermediate support.	3			3.5
Cables sheathed with plastics or having an oil-resisting and flame-retardant (hofr) sheath, in heavy gauge steel conduit of a diameter not less than 20 mm and not jointed in its span.	3	5.8 (for all types	5.2 (for all types	3
Bare or covered overhead lines supported by insulators without intermediate support.	30			3.5
Cables sheathed with plastics or having an oil-resisting and flame-retardant (hofr) sheath, supported by a catenary wire.	No limit			3.5
Overhead cable incorporating a catenary wire.	Subject to Item (xiv)			3.5
Bare or covered overhead lines installed in accordance with the Electricity Supply Regulations 1988 (as amended).	No limit			5.2

Notes:

(†) This column is not applicable in agricultural premises.

Note: In some special cases, such as in yacht marinas or where large cranes are present, it will be necessary to increase the height of span above ground over the minimum given in the Table.

TABLE G4
Spacings of supports for conduits

Nominal diameter of conduit	Maximum distance between supports					
	Rigid metal		Rigid insulating		Pliable	
	Horizontal	Vertical	Horizontal	Vertical	Horizontal	Vertical
mm	m	m	m	m	m	m
Up to 16	0.75	1.0	0.75	1.0	0.3	0.5
Over 16 and up to 25	1.75	2.0	1.5	1.75	0.4	0.6
Over 25 and up to 40	2.0	2.25	1.75	2.0	0.6	0.8
Over 40	2.25	2.5	2.0	2.0	0.8	1.0

Notes:

(1) The figures given above are the maximum spacings.

(2) Conduit boxes supporting luminaires or electrical accessories will require separate fixing.

(3) Plastic conduits will require closer spacing of fixings in areas of high ambient temperature, and note must be taken of the manufacturers declared maximum temperature.

(4) The spacings in the table allow for maximum fill of cables in compliance with Appendix A and thermal limits to the relevant British Standards. They assume that the conduit is not exposed to other mechanical stress.

(5) A flexible conduit is not normally required to be supported in its run. Supports should be positioned within 300 mm of bends or fittings.

TABLE G5
Spacings of supports for cable trunking

Cross-sectional area of trunking	Maximum distance between supports			
	Metal		Insulating	
	Horizontal	Vertical	Horizontal	Vertical
mm^2	m	m	m	m
Up to 700	0.75	1.0	0.5	0.5
Over 700 and up to 1500	1.25	1.5	0.5	0.5
Over 1500 and up to 2500	1.75	2.0	1.25	1.25
Over 2500 and up to 5000	3.0	3.0	1.5	2.0
Over 5000	3.0	3.0	1.75	2.0

Notes:

(1) The spacings in the table allow for maximum fill of cables in compliance with Appendix A and thermal limits to the relevant British Standards. They assume that the trunking is not exposed to other mechanical stress.

(2) The above figures do not apply to lighting suspension trunking, or where specially strengthened couplers are used.

(3) Plastic trunking will require closer spacing of fixings in areas of high ambient temperature, and note must be taken of the manufacturers declared maximum temperature.

Appendix H: Maximum demand and diversity

This Appendix gives some information on the determination of the maximum demand for an installation and includes the current demand to be assumed for commonly used equipment. It also includes some notes on the application of allowances for diversity.

The information and values given in this Appendix are intended only for guidance because it is impossible to specify the appropriate allowances for diversity for every type of installation and such allowances call for special knowledge and experience. The figures given in Table H2, therefore, may be increased or decreased as decided by the competent person responsible for the design of the installation concerned. For blocks of residential dwellings, large hotels, industrial and large commercial and office premises, the allowances are to be assessed by a competent person.

The current demand of a final circuit is determined by summating the current demands of all points of utilisation and equipment in the circuit and, where appropriate, making an allowance for diversity. Typical current demands to be used for this summation are given in Table H1. For the standard circuits using BS 1363 socket-outlets, detailed in Appendix C, an allowance for diversity has been taken into account and no further diversity should be applied.

The current demand of a distribution system or distribution circuit supplying a number of final circuits may be assessed by using the allowances for diversity given in Table H2 which are applied to the total current demand of all the equipment supplied by that circuit and not by summating the current demands of the individual final circuits obtained as outlined above. In Table H2 the allowances are expressed either as percentages of the current demand or, where followed by the letters f.l., as percentages of the rated full load current of the current-using equipment. The current demand for any final circuit which is a standard circuit arrangement complying with Appendix C is the rated current of the overcurrent protective device of that circuit.

An alternative method of assessing the current demand of a circuit supplying a number of final circuits is to summate the diversified current demands of the individual circuits and then apply a further allowance to diversity but in this method the allowances given in Table H2 are not to be used, the values to be chosen being the responsibility of the designer of the installation.

The use of other methods of determining maximum demand is not precluded where specified by a competent person.

After the design currents for all the circuits have been determined, enabling the conductor sizes to be chosen, it is necessary to check that the design complies with the requirements of Part 4 of BS 7671 and that the limitation on voltage drop is met.

TABLE H1
Current demand to be assumed for points of utilisation and current-using equipment

Point of utilisation or current-using equipment	Current demand to be assumed
Socket-outlets other than 2 A socket-outlets	Rated current
2 A socket-outlets	At least 0.5 A
Lighting outlet	Current equivalent to the connected load, with a minimum of 100 W per outlet
Final circuits for discharge lighting are arranged so as to be capable of carrying the total steady current of the lamp(s) and any associated gear and also their harmonic currents. Where more exact information is not available, the demand in volt amperes is taken as the rated lamp watts multiplied by not less than 1.8. This multiplier is based upon the assumption that the circuit is corrected to a power factor of not less than 0.85 lagging and takes into account controlgear losses and harmonic current.	
Electric clock, electric shaver supply unit (complying with BS 3535), shaver socket-outlet (complying with BS 4573), bell transformer, and current-using equipment of a rating not greater than 5 VA	May be neglected
All other stationary equipment	British Standard rated current, or normal current

TABLE H2
Allowances for diversity

Purpose of final circuit fed from conductors or switchgear to which diversity applies	Type of premises: Individual household installations including individual dwellings of a block	Small shops, stores, offices and business premises	Small hotels, boarding houses, guest houses, etc.
1. Lighting	66 % of total current demand	90 % of total current demand	75 % of total current demand
2. Heating and power (but see 3 to 8 below)	100 % of total current demand up to 10 amperes +50 % of any current demand in excess of 10 amperes	100 % f.l. of largest appliance +75 % f.l. of remaining appliances	100 % f.l. of largest appliance +80 % f.l. of 2nd largest appliance +60 % f.l. of remaining appliances
3. Cooking appliances	10 amperes +30 % f.l. of connected cooking appliances in excess of 10 amperes +5 amperes if socket-outlet incorporated in control unit	100 % f.l. of largest appliance +80 % f.l. of 2nd largest appliance +60 % f.l. of remaining appliances	100 % f.l. of largest appliance +80 % f.l. of 2nd largest appliance +60 % f.l. of remaining appliances
4. Motors (other than lift motors which are subject to special consideration	not applicable	100 % f.l. of largest motor +80 % f.l. of 2nd largest motor +60 % f.l. of remaining motors	100 % f.l. of largest motor +50 % f.l. of remaining motors
5. Water-heaters (instantaneous type)*	100 % f.l. of largest appliance +100 % f.l. of 2nd largest appliance +25 % f.l. of remaining appliances	100 % f.l. of largest appliance +100 % f.l. of 2nd largest appliance +25 % f.l. of remaining appliances	100 % f.l. of largest appliance +100 % f.l. of 2nd largest appliance +25 % f.l. of remaining appliances
6. Water-heaters (thermostatically controlled)		no diversity allowable†	
7. Floor warming installations		no diversity allowable†	

TABLE H2
Continued

Purpose of final circuit fed from conductors or switchgear to which diversity applies	Type of premises: Individual household installations including individual dwellings of a block	Small shops, stores, offices and business premises	Small hotels, boarding houses, guest houses, etc.
8. Thermal storage space heating installations		no diversity allowable†	
9. Standard arrangement of final circuits in accordance with Appendix E	100 % of current demand of largest circuit +40 % of current demand of every other circuit	100 % of current demand of largest circuit +50 % of current demand of every other circuit	
10. Socket-outlets other than those included in 9 above and stationary equipment other than those listed above	100 % of current demand of largest point of utilisation +40 % of current demand of every other point of utilisation	100 % of current demand of largest point of utilisation +75 % of current demand of every other point of utilisation	100 % of current demand of largest point of utilisation +75 % of current demand of every other point in main rooms (dining rooms, etc) +40 % of current demand of every other point of utilisation

Notes:

* For the purpose of this table an instantaneous water-heater is deemed to be a water-heater of any loading which heats water only while the tap is turned on and therefore uses electricity intermittently.

† It is important to ensure that the distribution boards etc. are of sufficient rating to take the total load connected to them without the application of any diversity.

Appendix I: Minimum separating distances between electricity supply cables and telecommunications or control cables

The effects of electromagnetic interference (EMI) on an installation and its equipment, have been a problem that many have experienced. Designers and installers faced with this phenomenon have, in the past, overcome the problems in a number of ways e.g. use of filtering devices, earthing and bonding, and segregation or separation of cables.

However, with the publication of a European Directive on EMC, it is now a requirement for manufacturers and installers to ensure that equipment, when installed, will:

(a) not interfere with other equipment in the vicinity (e.g. it must limit its emissions); and

(b) be able to withstand likely interference levels present in the vicinity, (e.g. it must have sufficient immunity).

Regulation 515-02 of BS 7671 outlines the requirements for emission of electromagnetic influences and interference from such influences, but does not give any guidance on how these should be achieved. Some installers are now proposing tests are carried out on completed installations.

To comply with the requirements of the Electromagnetic Compatability Regulations 1992 legislation only equipment manufactured so as to comply with the relevant standards for protection against interference and, where available for immunity is used, e.g.:

BS EN 55011 for Industrial, scientific and medical equipment.

BS EN 55015 for Electrical lighting and similar.

BS EN 55022 (and prEN 55024) for Information technology equipment.

Where equipment does not have a specific standard, or where a specific standard covers only emissions and not immunity, equipment should conform to the appropriate generic standard:

For domestic, commercial and light industrial equipment these are:

BS EN 50081-1 for emissions; and BS EN 50082-1 for immunity.

For heavy industrial equipment:

BS EN 50081-2 for emissions; and BS EN 50082-2 for immunity.

Further information is also available in British Standards on segregation of cables e.g. BS 6701 : 1994 (as amended) for the installation of telecommunication cables.

The avoidance of problems should be given due regard at the design stage and segregation is considered to be a way of overcoming the problems of EMI.

The segregation details given in Table I1 apply only to cases where adjacent runs of both power and signal cables are envisaged.

With regard to the figures in the table, it is important to note that these are minimum distances and it is assumed the equipment connected to the cable meets the Generic Standards, which may well not be the case for existing equipment. Moreover, the manufacturers of electronic equipment may, in addition to any requirements for cable separation on the grounds of safety, specify that greater separations are used for installations including their equipment.

A major factor affecting the performance of the cables in EMC terms is the quality of termination the screen. It is therefore important that the correct termination is used.

The figures apply to up to 100 m parallel run and with up to 125 A flowing in the power cable.

TABLE I1
Example of power cable and information technology cables that run parallel with a metallic divider.

	Without divider or non-metallic divider (1)	Aluminium divider	Steel divider
Unscreened power cable and IT cable	200 mm	100 mm	50 mm
Unscreened power cable and screened IT cable	50 mm	20 mm	5 mm
Screened power cable and unscreened IT cable	30 mm	10 mm	2 mm
Screened power cable and screened IT cable (2)	0	0	0
Note 1: It is assumed that in case of a metallic divider, the design of the CMS will achieve a screening attenuation related to the material used for the divider. Note 2: The screened IT cables shall comply with EN 50288 series.			

Therefore, for horizontal cabling the following applies:

- If the full length is no greater than 35 m no separation is required.
- For lengths greater than 35 m the separation distances apply to the full length excluding the final 15 m attached to the outlet.

The second amendment to BS 7671 : 1992, AMD 9781 deleted circuit categories and substituted "Voltage Bands". Band I covers effectively ELV circuits (including SELV), and Band II effectively covers LV circuits. These bands only relate to separation for electrical safety and do not relate to separation for electromagnetic interference requirements. It is still possible in some cases, to have a cable for a 230 V power circuit near to a cable for an intruder alarm. This method of installation would be unlikely to meet the requirements of the legislation, and interference on the security system could be experienced.

The need for system cables to be separated is in conflict with the usual building design for the minimum space to be allowed for cable routes but generally, separation between system cables should be as large as possible, but available space will always be a limiting requirement. Especially sensitive systems or systems that will emit interference should be identified and adequate provisions made in the design with the advice of the system manufacturer. It may be that building designers will have to allow more space for electrical cable routes and risers than they would wish. All cable systems should be identified as necessary for ease of future modification and maintenance.

High voltage system cables should be segregated from cables of other systems and clearly identified for a general safety precaution, as well as for EMC reasons. If separation cannot be achieved some form of protective screen or barrier may be required. Table I2 recommends minimum separation distances between systems.

TABLE I2
Separation of systems

Voltage to earth	Normal separation distances	Exceptions to normal separation distances, plus conditions to exception
Exceeding 600 V a.c. and 900 V d.c.	150 mm	50 mm, as long as a divider which maintains separation of 50 mm is inserted between two sets of cables. The divider should be made of a rigid, non-conducting, non-flammable material

Appendix J: Permitted leakage currents

Equipment manufacturing standards allow a specific maximum leakage current for types of equipment. Additionally, certain equipment produces earth leakage current as part of its normal operation. BS 7671 Section 607 gives requirements relating to electrical equipment with high earth leakage current (normally exceeding 3.5 mA).

TABLE J1
Earth leakage current limits

	Household appliance BS 3456 Clause 16.2 (See also BS EN 60335)	Luminaires BS 4533 Clause 10.3 (See also BS EN 60598-1	Hand-held tools BS 2769 Clause 12	Information technology equipment BS EN 60950 Clause 5.2	Catering equipment BS 5784
	mA	mA	mA	mA	mA
For Class 0, Class 0I and Class III appliances	0.5		0.5	N/A	0.5
For portable and hand-held Class I equipment	0.75	1.0	0.75	0.75	0.75
For other stationary Class I appliances	0.75 or 0.75 per kW rated input of the appliance, whichever is the greater with a maximum of 5 mA (Note 1)	N/A	N/A	3.5 (Note 3)	3.5 (Note 3)
For Class II appliances	0.25 (Note 2)	0.5	0.25	0.25	0.25

(*Notes:* See next page)

Notes toTable J1 :

(1) For stationary Class I appliances with heating elements which are detachable or can be switched off separately, 0.75 mA or 0.75 mA per kW rated input for each element or group of elements, whichever is the greater, with a maximum of 5 mA for the appliance as a whole.

(2) Between live parts and metal parts of Class II appliances separated from live parts by basic insulation only, if the appliance is classified according to degree of protection against moisture as:

- ordinary appliance..5.0 mA
- other than ordinary appliance.............................3.5 mA

The values specified above are doubled:

- if the appliance has no control device other than a thermal cut-out, a thermostat without an 'off' position or an energy regulator without an 'off' position

- if all control devices have an 'off' position with a contact opening of at least 3 mm and disconnection in each pole.

However, for Class II appliances, doubling of the value of 0.25 mA is only allowed if all control devices have an 'off' position with a contact opening of at least 3 mm and all-pole disconnection.

(3) Class I stationary equipment which is permanently connected by an industrial plug and socket to BS 4343 (BS EN 60309-2), or with an earth leakage current exceeding 3.5 mA, shall be subject to the following conditions:

- leakage current shall not exceed 5 % of the input current per phase. Where the load is unbalanced the largest of the three-phase currents shall be used for this calculation. Where necessary, the tests in Sub-clause 5.23 and 5.24 shall be used but with a measuring instrument of negligible impedance;

- the cross-sectional area of the internal protective earthing conductor shall be not less than 1.0 mm^2 in the path of high leakage current;

- a label bearing the following warning, or similar wording, shall be affixed adjacent to the equipment primary power connection:

HIGH LEAKAGE CURRENT
Earth connection essential before connecting supply

(4) Up to 1 kVA rated input: 1.0 mA increasing by 1.0 mA/kVA to a maximum of 5.0 mA.

(5) For Class I appliances that are intended to be connected with a flexible cable or cord and plug-top a maximum of 1 mA, or 1 mA per kW of rated input up to a maximum of 10 mA. This maximum does not apply to appliances connected directly to the fixed wiring of the electrical installation.

Appendix K: Standard symbols and graphical symbols for general electrical purposes

BS 3939: Graphical symbols for electrical power, telecommunications and electronics diagrams has been replaced by BS EN 60617 however, the two standards are identical. Installation layout symbols are contained in BS EN 60617-11: 1997. The following symbols are extracted mainly from BS EN 60617, supplemented by references from other Standards.

A. General symbols

V .... volts

A .... amperes

Hz .... hertz

W .... watts

kW .... kilowatts

F .... farads

p.u. .... per unit

ph .... phase

p.f. .... power factor

L .... line

N .... neutral

h .... hours

min .... minutes

s .... seconds

⎓ (or d.c.) .... direct current

∼ (or a.c.) .... alternating current

2∼ .... two-phase alternating current

2N ∼.. two-phase alternating current with neutral

3∼ .... three-phase alternating current

3N∼ .... three-phase alternating current with neutral

IPXX .... IP number (see Appendix B)

Fault (indication of assumed fault location).

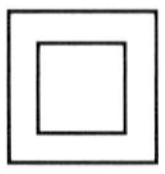

Class II appliance (Equipment in which protection against electric shock does not rely on basic insulation only, but in which additional safety precautions such as supplementary insulation are provided, there being no provision for the connection of exposed metalwork of the equipment to a protective conductor, and no reliance upon precautions to be taken in the fixed wiring of the installation (see BS 2754)).

Class III appliance (Equipment in which protection against electric shock relies on supply at SELV and in which voltages higher than those of SELV are not generated (see BS 2754)).

Safety isolating transformer. Class III equipment must be supplied from a safety isolating transformer to BS 3535. The safety isolating transformer will have this identifying mark upon it.

Isolating transformer.

Protective earth, general symbol (preferred to ⏚).

In the International System of units (known as SI), there are Seven base units as shown below, other quantities are derived from these.

Quantity	Name of base unit	Unit symbol
Length	metre	m
Mass	kilogram	kg
Time	second	s
Electric current	ampere	A
Thermodynamic temperature	kelvin	K
Amount of substance	mole	mol
Luminous intensity	candela	cd

Multiples and sub-multiples of quantities:

10^{18}	exa	E				10^{-3}	milli	m
10^{15}	peta	P	10^{2}	hecto	h	10^{-6}	micro	μ
10^{12}	tera	T	10^{1}	deca	da	10^{-9}	nano	n
10^{9}	giga	G	10^{-1}	deci	d	10^{-12}	pico	p
10^{6}	mega	M	10^{-2}	centi	c	10^{-15}	femto	f
10^{3}	kilo	k				10^{-18}	atto	a

Powers in steps of 3 are preferred, but some others have common usage (e.g. centimetre, cm; decibel dB).

SI Derived units

The units of all physical quantities are derived from the base and supplementary SI units, and certain of them have been named. These, together with some common compound units, are given here:

Quantity	Unit name	SI Units	Unit symbol
Force	newton	kg m/s^2	N
Energy	joule	N m	J
Power	watt	J/s	W
Pressure, stress	pascal	N/m^2	Pa
Electric potential	volt	J/C, W/A	V
Electric charge, electric flux	coulomb	A s	C
Magnetic flux	weber	V s	Wb
Magnetic flux density	tesla	Wb/m^2	T
Resistance	ohm	V/A	
Conductance	siemens	A/V	S
Capacitance	farad	C/V	F
Inductance	henry	Wb/A	H
Celsius temperature	degree celsius	K	°C
Frequency	hertz	s^{-1}	Hz
Luminous flux	lumen	cd sr	lm
Illuminance	lux	lm/m^2	lx
Activity (radiation)	becquerel	s^{-1}	Bq
Absorbed dose	gray	J/kg	Gy
Dose equivalent	sievert	J/kg	Sv

Mass density	kilogram per cubic metre	kg/m^3
Torque	newton metre	N m
Electric field strength	volt per metre	V/m
Magnetic field strength	ampere per metre	A/m
Thermal conductivity	watt per metre kelvin	Wm^{-1}K^{-1}
Luminance	candela per square metre	cd/m^2

B. Symbols for use in schematic wiring diagrams

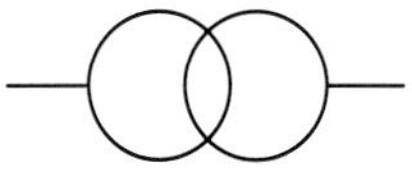
Transformer - general symbol

General wiring, with joint

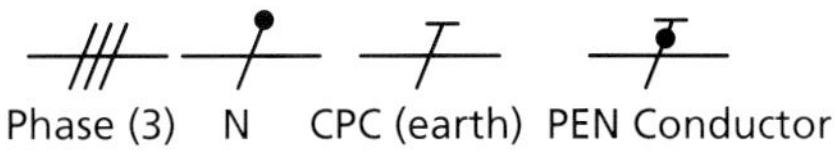

Identification of specific conductors

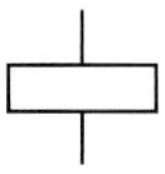
Operating device, general symbol (coil)

Make contact - normally open

Break contact - normally closed

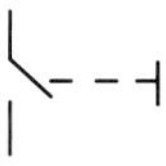
Manually operated switch, general symbol

Three-phase winding - delta

Three-phase winding - Star

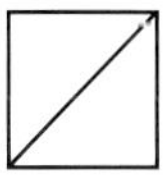
Changer, general symbol
Converter, general symbol

Notes:

(1) If the direction of change is not obvious, it may be indicated by an arrowhead on the outline of the symbol.

(2) A symbol or legend indicating the input or output quantity, waveform etc. may be inserted in each half of the general symbol to show the nature of the change.

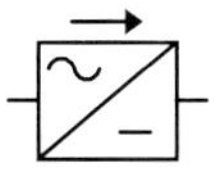
Rectifier

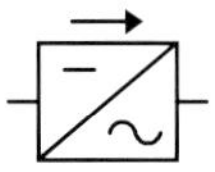
Invertor

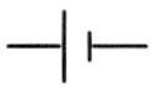
Primary cell - longer line positive, shorter line negative

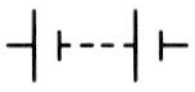
Battery

fuse link, rated current in amperes

Making and breaking current:

 switch

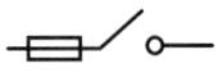 switch-fuse

 fuse-switch

Isolating:

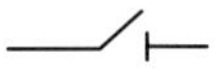 Isolator (Disconnector), general symbol

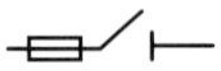 Disconnector - fuse (fuse combination unit)

 Fuse - disconnector

Making, breaking and isolating:

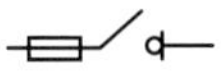 Switch - disconnector

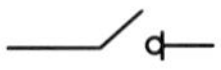 Switch - disconnector - fuse (fuse combination unit)

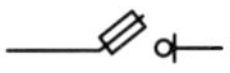 Fuse - switch - disconnector

 Capacitor, general symbol

 Inductor, coil, winding or choke

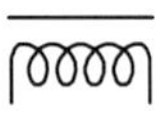 Inductor, coil, winding or choke with magnetic core

 Semi Conductor Diode - general symbol

 Microphone

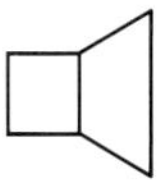

Loudspeaker

Antenna, general symbol

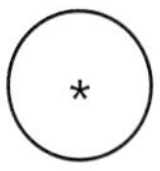

Machine, general symbol
* Function M = Motor G = Generator

Generator, general symbol

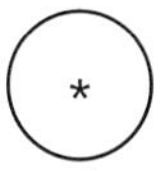

Indicating instrument, general symbol
* function V = Voltmeter A = Ammeter etc

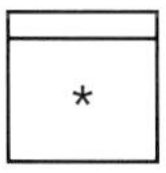

Integrating instrument or Energy meter
* function Wh = Watt-hour
VArh = Volt ampere reactive hour

Lamp, or signal lamp, general symbol

C. Location symbols for installations

Machine, general symbol
* function etc

Load, general symbol
* details

Motor starter, general symbol
* indicates type etc

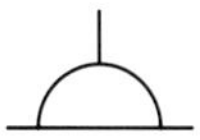

Socket-outlet, general symbol

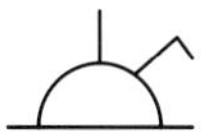

Switched socket-outlet

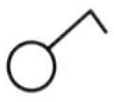

Switch, general symbol

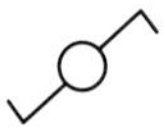

2 way switch, single-pole

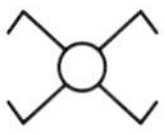

Intermediate switch

 Pull switch, single-pole

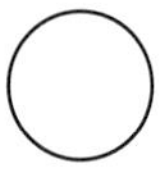 Lighting outlet position - general symbol

 Fluorescent luminaire, general symbol

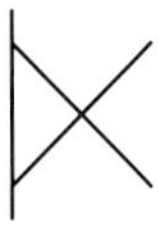 Wall mounted luminaire

 Emergency lighting luminaire (or special circuit)

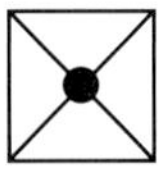 Self-contained emergency lighting luminaire

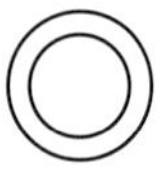 Push button

 Clock, general symbol

 Bell

 Buzzer

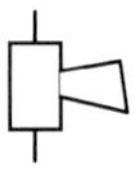 Horn

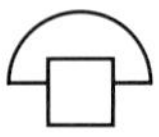 Telephone handset, general symbol

Appendix L: Addresses of associated bodies and identification symbols

Amalgamated Engineering and Electrical Union (AEEU)
Hayes Court
West Common Road
Bromley BR2 7AU

Tel: 020 8462 7755
Fax: 020 8315

Association of Consulting Engineers (ACE)
Alliance House
12 Caxton Street
London SW1H 0QL

Tel: 020 8222 6557
Fax: 020 8222 0750

Association of Manufacturers of Domestic Appliances (AMDEA)
Rapier House
40-46 Lamb's Conduit Street
London WC1N 3NW

Tel: 020 7405 0666
Fax: 020 7405 6609

Association of Short Circuit Testing Authorities (ASTA)
ASTA House
Chestnut Field
Rugby
Warwickshire CV21 2TL

Tel: 01788 578435
Fax: 01788 573605

British Approvals Service for Cables (BASEC)
23 Presley Way
Crownhill
Milton Keynes
Buckinghamshire MK8 0ES

Tel: 01908 267300
Fax: 01908 267255

British Approval Services for Electrical Equipment in Flammable Atmospheres (BASEEFA)
(See Electrical Equipment Certification Services)

British Cable Manufacturers' Confederation (BCMC)
Cable House
56 Palace Road
East Molesey
Surrey KT8 9DW

Tel: 020 8941 4079
Fax: 020 8783 0104

British Electrical Systems Association (BESA)
Granville Chambers
2 Radford Street
Stone
Staffordshire ST15 8DA

Tel: 01785 812426
Fax: 01785 818157

British Electrotechnical Allied Manufacturers Association (BEAMA)
Westminster Tower
Albert Embankment
London SE1 7SL

Tel: 020 7793 3000
Fax: 020 7793 3003

British Electrotechnical Approvals Board (BEAB)
1 Station View
Guildford
Surrey GU1 4JY

Tel: 01483 455 466
Fax: 01483 455 477

British Radio and Electronic Equipment Manufacturers Association (BREMA)
Landseer House
19 Charing Cross Road
London WC2H 0ES

Tel: 020 7930 3206
Fax: 020 7839 4613

British Standards Institution (BSI)
(also BSI Technical Help for Exporters)
389 Chiswick High Road
London W4 4AL

Tel: 020 8996 9000
Fax: 020 8996 7400

Building Services Research and Information Association (BSRIA)
Old Bracknell Lane West, Bracknell
Berkshire RG12 7AH

Tel: 01344 426511
Fax: 01344 487575

Chartered Institution of Building
Services Engineers (CIBSE)
222 Balham High Road
London SW12 9BS

Tel: 020 8675 5211
Fax: 020 8675 5449

City and Guilds (C & G)
76 Portland Place
London W1N 4AA

Tel: 0207 278 2468
Fax: 0207 278 9460

Copper Development Association
Verulam Industrial Estate
224 London Road
St Albans
Hertfordshire AL1 1AQ

Tel: 01727 731200 (Technical Helpline)
Fax: 01727 731216

Department of Trade and Industry (DTI)
151 Buckingham Palace Road
London SW1W 9SS

Tel: 020 7215 5000

Electricity Association (EA)
30 Millbank
London SW1P 4RD

Tel: 020 7963 5700
Fax: 020 7963 5959

Electrical Contractors' Association (ECA)
ESCA House
34 Palace Court
Bayswater
London W2 4HY

Tel: 020 7313 4800
Fax: 020 7221 7344

Electrical Contractors' Association of
Scotland trading as SELECT
Bush House
Bush Estate
Midlothian EH26 0SB

Tel: 0131 445 5577
Fax: 0131 445 5548

Electrical Equipment Certification
Services (Formally BASEEFA)
Harpur Hill
Buxton
Derbyshire SK17 9JN

Tel: 01298 28000
Fax: 01298 28244

Electrical Installation Equipment Manufacturers'
Association Ltd (EIEMA)
Westminster Tower
3 Albert Embankment
London SE1 7SL

Tel: 020 7793 3013
Fax: 020 7735 4158

Energy Industries Council (EIC)
Newcombe House
45 Notting Hill Gate
London
W11 3LQ

Tel: 020 7221 2043
Fax: 020 7221 8813

Engineering Equipment and Material Users
Association (The) (EEMUA)
14-15 Belgrave Square
London SW1X 8PS

Tel: 020 7235 5316/7
Fax: 020 7245 8937

ERA Technology Ltd
Cleeve Road
Leatherhead
Surrey KT22 7SA

Tel: 01372 374151
Fax: 01372 367099

Federation of the Electronics Industry (FEI)
Russell Square House
10-12 Russell Square
London WC1B 5EE

Tel: 0207 331 2000
Fax: 0207 331 2040

Fibreoptic Industry Association (FIA)
10-15 The Arcade Chambers
High Street
Eltham
London SE9 1BG

Tel: 0208 959 6617 and 0208 850 5614

GAMBICA
(Association for the Instrumentation, Control and Automation Industry in the UK)
Westminster Tower
3 Albert Embankment
London SE1 7SW

Tel: 0207 793 3050
Fax: 0207 793 7635

Health and Safety Executive (HSE)
Rose Court
2 Southwark Bridge
Southwark
London SE1 9HS

Tel: 020 7717 6000
Fax: 020 7717 6717

Health and Safety Information Line Tel: 0541 545500

Health and Safety Executive
Library and Information Service
Health and Safety Laboratories
Broad Lane
Sheffield S3 7HQ

Tel: 01142 892345
Fax: 01142 892333

HSE Books
Sudbury
Suffolk

Tel: 01787 881165

Institution of Electrical Engineers (IEE)
Savoy Place
London WC2R 0BL

Tel: 020 7240 1871
Fax: 020 7497 2143
Technical Enquiries Tel: 020 7344 8420

Institution of Incorporated Engineers (IIE)
Savoy Hill House
Savoy Hill
London WC2R 0BS

Tel: 020 7836 3357
Fax: 020 7497 9006

Joint Industry Board for the
Electrical Contracting Industry(JIB)
Kingswood House
47-51 Sidcup Hill
Sidcup
Kent DA11 9HP

Tel: 020 8302 0031
Fax: 020 8309 1103

Lighting Association (LA)
Stafford Park 7
Telford
Shropshire TF3 3BD

Tel: 01952 290905
Fax: 01952 290906

London District Surveyors Association
PO Box 15
London SW17 7BQ

Lighting Industry Federation (LIF)
207 Balham High Road
London SW17 7BQ

Tel: 020 8675 5432
Fax: 020 8673 5880

National Electrotechnical Training (NET)
34 Palace Court
London
W2 4HY

Tel: 020 7313 4846
Fax: 020 7221 7344

NHBC (National House-Building Council)
Buildmark House
Chiltern Avenue
Amersham
Buckinghamshire HP6 5AP

Tel: 01494 434477
Fax: 01494 728521

National Inspection Council for Electrical Installation
Contracting (NICEIC)
Vintage House
37 Albert Embankment
London SE1 7UJ

Tel: 020 7564 2323
Fax: 020 7564 2370

National Joint Utilities Group (NJUG)
30 Millbank
London SW1P 4RD

Tel: 020 7963 5720

Royal Institute of British Architects (RIBA)
66 Portland Place
London W1N 4AD

Tel: 020 7580 5533
Fax: 020 7255 1541

SELECT - Trading name of Electrical
Contractors Association of Scotland
Bush House
Bush Estate
Midlothian EH26 0SB

Tel: 0131 445 5577
Fax: 0131 445 5548

Telecommunications Industry Association (TIA)
20 Drakes Mews
Crownhill
Milton Keynes MK8 0ER

Tel: 01908 265090
Fax: 01908 263852

Identification Symbols

British Approvals Board for Telecommunications

BABT is the UK Notified Body for the TTE Directive and the UK regulatory body responsible for the approval of terminal equipment for connection to UK and European public networks

British Electrotechnical Approvals Board

BEAB is the UK National Certification Body for domestic and light commercial electrical equipment

IEC International Certification Agreement (IECEE/CB) and the Cenelec Certification Agreement (CCA)

The two schemes facilitate international trade in electrical equipment through mutual acceptance of reports and test certificates. They allow approval marks relating to electrical safety to be gained world-wide, normally without further testing.

CE marking denotes that a product conforms to certain New Approach EC Directives designed to remove barriers to trade within the European Union. Before the product can be offered for sale anywhere within the EU it must conform to the requirements of all the Directives that apply to the product

European Telecommunications Standards Institute

ETSI is the European body responsible for the generation of standards for equipment to be attached to, or used within, telecommunications networks, including radio equipment. These include voluntary test standards, (ETSs) as well as those forming the basis of regulatory requirements (TBRs) for the TTE Directive.

Federation Communications Commission

The FCC is the independent government agency in the USA responsible for regulating all forms of communications and use of the radio spectrum, by the general public in the USA.

National Accreditation for Measurement and Sampling

NAMAS accreditations of test and calibration laboratories are granted by the United Kingdom Accreditation Service (UKAS). NAMAS accreditation is a recognition that the test procedures and quality systems operated by a laboratory meet required standards, in particular those of BS 45001.

British Approval Services for Cables (BASEC)

Underwriters Laboratories

(UL) is the largest of the USA's nationally recognised testing laboratories, offering testing and certification (Listing) to established national and International standards of safety.

Voluntary Control Council for Interference by Information Technology Equipment

The VCCI EMC compliance programme applies to the Information Technology Equipment (ITE) to be supplied for the Japanese market and although it is voluntary, it is widely supported by not only major Japanese companies but overseas companies as well.

Production Conformity Certification Product Approval

Product Conformity Certification Product Approval

Product Conformity Certification involves an assessment of the product type against the requirements of a National Standard and an assessment of the manufacturer's Quality Management System against BS 5750 or equivalent. Where a product type is outside the scope of a National Standard due to for example special design features ASTA will produce an ASTA Standard for the purposes of Product Approval. Such a Standard will make use of the requirements in National Standards wherever possible.

Certification of Supplier Quality Management Systems

Certification of Supplier Quality Management Systems

The Scheme which is open to any manufacturer provides third party certification of a manufacturer's Quality Management System to BS 5750: Part 1, 2 or 3. (ISO 9001, ISO 9002 or ISO 9003). On request, the scheme can be extended to cover any other national or international management system quality Standard. The Scheme was developed initially for single firm application and, as such, does not depend upon the production and general release of assessment or technical schedules. Any special requirements, unique to an Applicant and his defined scope, are prescribed in quality plans called up by the manufacturer's Quality Manual and related working procedural documents. Provision has been made within the Scheme for special assessment schedules to be produced and published should this be considered desirable by a particular industrial sector.

Index